MyMathLab®
Notebook

Gex, Incorporated

Developmental Mathematics

John Squires

Chattanooga State Community College

Karen Wyrick

Cleveland State

PEARSON

Boston San Francisco New York
London Toronto Sydney Tokyo Singapore Madrid
Mexico City Munich Paris Cape Town Hong Kong Montreal

ISBN-13: 978-0-321-78563-3
ISBN-10: 0-321-78563-0

ISBN-13: 978-0-321-75883-5
ISBN-10: 0-321-75883-8

9 10 11 12 EBM 16 15 14 13

www.pearsonhighered.com

MyMathLab Developmental Mathematics: A Modular Approach

John Squires, Karen Wyrick

Table of Contents

Whole Numbers
Topic 1.1 Whole Numbers

Vocabulary
whole number   •   standard form   •   place value   •   expanded notation

1.   To better understand the value of each digit in a number, we can write it in

_____.

Step-by-Step Video Notes
Watch the Step-by-Step Video lesson and complete the examples below.

Example	Notes
1. Consider the number 13,579. What digit is the tens place value? 7 What is the place value of the 3? _____ What is the actual value of the 3? 3000	
2. Write 4295 in expanded form. The 4 is in the thousands place value, so the actual value is 4000. What is the actual value of the 2? □ What is the actual value of the 9? □ What is the actual value of the 5? □ $4295 = 4000 + \square00 + \square0 + 5$ Answer:	

Example	Notes
4. Convert 1209 to words. Write the number in the first period on the left followed by the period name and a comma. one _____ , Do this for the next period, but remember that the "ones" do not need the name of the period. Answer:	
6. Write twenty-four thousand, seven hundred twelve in standard form. Read the number from left to right. Write the number in the first period followed by a comma. ☐ , Continue. Answer:	

Helpful Hints
When converting numbers to words or writing numbers in standard form, you always start at the left.

The word "and" is not included when reading or writing numbers.

Concept Check
1. How many place values are in a period?

Practice

Convert to words.

2. 5287

3. 321,608

Write in standard form.

4. two thousand six hundred four

5. twelve thousand, forty-eight

2

Whole Numbers
Topic 1.2 Rounding

Vocabulary

estimating   •   rounding   •   rounding down   •   rounding up

1. _____ is finding a number close to the exact number, but easier to work with.

Step-by-Step Video Notes
Watch the Step-by-Step Video lesson and complete the examples below.

Example	Notes
3. Round 506,243 to the nearest thousand. Underline the digit in the place value to which you are rounding. 50<u>6</u>,243 What is the digit to the right of the underlined digit? ☐ It is 4 or less, so the underlined digit stays the same. Replace the digit to the right of 6 with zeros. Answer: 506,☐☐☐	
4. Round 101,697 to the nearest hundred. 101,<u>6</u>97 Answer:	

Example	Notes
5. Round 1296 to the nearest ten. 12<u>9</u>6 Look at ☐, which is 5 or more so the under Lined digit increases by one. Note that when a 9 becomes a 10, the digit to the left of 9 also increases by one, so the 2 becomes a three. Answer: 13☐☐	
6. Round 449,985 to the nearest hundred. Answer:	

Helpful Hints
Remember to replace all the digits to the right of the digit you are rounding with zeros after rounding up or rounding down.

When a 9 is rounded up to ten, the digit to the left of the 9 must also be increased by one.

Concept Check
1. When rounding a number to the nearest place value, and the number to the right of the
    digit being rounded is 5, do you round up or round down?

Practice
Round the following numbers to the nearest place value indicated.
2. 4528 to the nearest ten

4. 894 to the nearest hundred

3. 23,179 to the nearest thousand

5. 49,926 to the nearest hundred

Whole Numbers
Topic 1.3 Adding Whole Numbers; Estimation

Vocabulary
addition   •   sum   •   estimation   •   commutative property of addition   •
addition property of zero   •   rounding   •   associative property of addition

1.   _____ occurs when you combine numbers.

2.   The property which states that changing the order when adding numbers does not change
     the sum is the _____.

Step-by-Step Video Notes
Watch the Step-by-Step Video lesson and complete the examples below.

Example	Notes
1. Find the sum of $5+6$. The sum is the answer to the addition, which is combining 5 and 6 and counting the total number of items. Answer:	
4. The statement $(3+5)+7 = 3+(5+7)$ is true. This shows that changing the grouping when adding numbers does not change the sum. Which property states this? Answer:	
7. Find the sum of $126+12$. Stack the numbers so that the digits in the same place value are lined up vertically. Begin at the right adding digits. $\begin{array}{r}126\\+12\\\hline\end{array}$ Answer:	

5

Example	Notes

9. Estimate by rounding to the nearest hundred and comparing to the exact answer.

$7026 + 13,479$

Round 7026 to the nearest hundred.   7☐00

Round 13,479 to the nearest hundred.

13,☐00

Find the sum of the rounded numbers. ☐

What is the exact answer? ☐

Answer:

Helpful Hints

When adding numbers vertically, if the sum of any column is more than 9, you need to carry the tens place value number to the next column.

Estimation is a process that can help you check that your exact answer is close to the actual answer.

Concept Check

1. Can you name three words or phrases that indicate addition?

Practice

Find the following.

2. $135 + 22$

4. 13 more than 8

3. $111 + 246 + 3189$

5. sum of 24 and 57

Whole Numbers
Topic 1.4 Subtracting Whole Numbers

Vocabulary
difference   •   sum   •   estimation   •   subtraction

1. _____ of two numbers is taking away one number or quantity from another.

2. The _____ is the answer to a subtraction problem.

Step-by-Step Video Notes
Watch the Step-by-Step Video lesson and complete the examples below.

Example	Notes
3. Subtract $10-4$. Take 4 away from 10. $10-4=\square$ Check by adding. $4+\square=10$ Answer:	
5. Subtract $57-38$. Begin with the ones place and subtract the bottom digit from the top. Since the top digit is smaller than the bottom digit, borrow from the next place value.$\quad \begin{array}{r}57\\-38\\\hline\end{array}$ Now subtract.$\quad \begin{array}{r}^{4\ 17}\\\cancel{5}\ \cancel{7}\\-3\ 8\\\hline\square\end{array}$ Check by adding. Answer:	

Example	Notes
6. Subtract $232 - 141$. Check by adding. Answer:	
8. Estimate by rounding to the nearest thousand and comparing to the exact answer. $11,976 - 1245$ Round 11,976 to the nearest thousand. ☐ Round 1245 to the nearest thousand. ☐ Find the difference of the rounded numbers. Compare to the exact answer. Answer:	

Helpful Hints
Any number minus itself is zero.

Estimation is a process that can help you check that your exact answer is close.

Concept Check
1. Can you name three words or phrases that indicate subtraction?

Practice
Find the following.

2. $86 - 28$

3. $3045 - 2824$

4. 13 decreased by 8

5. 24 less than 52

Whole Numbers
Topic 1.5 Basic Problem Solving

Vocabulary

addition   •   perimeter   •   problem solving   •   translating

1. To find the _____ of a figure, add the lengths of all its sides.

2. The procedure for _____ involves creating a plan in symbols or words and performing calculations.

Step-by-Step Video Notes
Watch the Step-by-Step Video lesson and complete the examples below.

Example	Notes
1. Translate the sum of 5 and 3 to symbols. Enter the operation indicated by the word sum. 5☐3 Simplify. Answer:	
3. A rectangular garden measures 9 feet in length and is 5 feet wide. How many feet of fencing are needed to enclose the garden? Understand the problem. We are trying to find out how much _____ is needed for the garden. Create a plan. We need to find the _____ of the lengths of the four sides. Find the answer. Check. Answer:	

Example	Notes
4. A local community college requires 64 credit hours for an Associate Degree. Kylie earned 28 credits this past year. How many more credit hours does Kylie need in order to get her degree? Enter the operation indicated from the key word or phrase in the problem. credit hours required ☐ credit hours earned is credit hours needed Answer:	
5. On Monday, Alex opened a checking account With an initial deposit of $300. She bought groceries for $75, spent $25 on gas, and spent $20 on new clothes. How much money is in her account after these purchases? Answer:	

Helpful Hints

When solving a problem, there are key words or phrases which can be translated into an operation.

After obtaining an answer from your calculation, make sure to check that this answers the question which was asked in the problem.

Concept Check

1. What two steps need to be done before calculation in solving a problem?

Practice

Today is Dan's turn to bring water to soccer practice. Dan's mom put 24 water bottles in the cooler and his dad put 6 water bottles to the cooler. Dan's soccer team has 19 players, but 3 are not at practice today. Each player takes one water bottle at practice. Determine the following values.

2. The number of water bottles in cooler before practice.

3. The number of players at practice today.

4. The number of water bottles taken at practice.

5. The number of water bottles left in cooler after practice.

Whole Numbers
Topic 1.6 Multiplying Whole Numbers

Vocabulary
product • factors • addition • multiplication property of one • multiplication
commutative property of multiplication   •   associative property of multiplication
multiplication property of zero • distributive property of multiplication over addition

1. The _____ is the answer to a multiplication problem.

2. The _____ states that the changing the grouping when multiplying numbers
   does not change the product.

Step-by-Step Video Notes
Watch the Step-by-Step Video lesson and complete the examples below.

Example	**Notes**
3. For 3×4, identify the factors. 3, $\square$ Find the product. $3 \times 4 = 3 + 3 + 3 + 3 = \square$ Answer:	
4. Rewrite $8(3+6)$ using the Distributive Property. $8(3) + \square(6)$ Simplify. Answer:	
7. Which property tells us that the following is true? $6 \cdot 25 = 25 \cdot 6$ Commutative Property, Associative Property, Distributive Property, Multiplication Property of Zero, Multiplication Property of One Answer:	

Example	Notes
12. Find the product $3 \cdot 248$.	

Multiply the bottom number by each digit on the top starting on the right.

$3 \times 8 = 24$

The 4 is written as the ones digit answer and the 2 is carried to the next place value.

$$\begin{array}{r} 2 \\ 248 \\ \underline{\times 3} \\ 4 \end{array}$$

Complete the multiplication.

Answer:

Helpful Hints

The Distributive Property can be used to make mental calculations easier.

When number is carried to the next place value in a multiplication problem, this number is added to the product of the next multiplication.

Concept Check

1. Can you name two words or phrases that indicate multiplication?

Practice

Rewrite using the Distributive Property. Simplify.

2. $3(9+7)$

Find the product.

4. $4 \cdot 164$

3. $7(5+2)$

5. $3 \cdot 192$

Whole Numbers
Topic 1.7 Dividing Whole Numbers

Vocabulary
division • quotient • dividend • divisor • long division • remainder
• divides exactly

1. The number you are dividing by is the _____.

2. The _____ is the answer to a division problem.

Step-by-Step Video Notes
Watch the Step-by-Step Video lesson and complete the examples below.

Example	Notes
2. Divide the following. $36 \div 4$ $4 \cdot \square = 36$, so $36 \div 4 = \square$ Answer:	
3. Divide the following. Use long division. $92 \div 4$ $\begin{array}{r} 2\,\square \\ 4{\overline{\smash{\big)}\,9\;2}} \\ \underline{-8} \\ \square\,2 \\ \underline{-\square\,2} \\ 0 \end{array}$ Answer:	

Example	Notes
4. Divide the following. Use long division. $3\overline{)148}$ $\square\square\,r\square$ $3\overline{)1\ 4\ 8}$ $-\square\square$ $\quad\square\,8$ $\ -\square\square$ $\qquad\square$ Answer:	
5. Find the quotient of 21 and 7. $7\cdot\square=21$, so $21\div7=\square$ Answer:	

Helpful Hints

In order to divide, you need to have mastered your multiplication facts. You can check any division problem using multiplication.

Remember that any non-zero number divided by itself is 1, and any number divided by 1 is the number itself. Zero divided by any non-zero number is 0, and division by 0 is undefined.

When translating for division, be careful to write the numbers in the correct order.

Concept Check
1. When using the long division symbol, which number goes inside? Which goes outside?

Practice

Divide the following.
2. $48\div6$

3. $119\div4$

Find the quotient of the following.
4. 32 and 8

5. 84 and 12

Whole Numbers
Topic 1.8 More with Multiplying and Dividing

Vocabulary
division • quotient • estimation • remainder • product • multiplication

1. The _____ is the answer to a multiplication problem.

Step-by-Step Video Notes
Watch the Step-by-Step Video lesson and complete the examples below.

Example	Notes
2. Calculate the following product. $32 \cdot 125$ Set up the multiplication. It is preferable to put the larger number on top when setting up this problem. $\begin{array}{r} 1\ 2\ 5 \\ \times\quad 3\ 2 \\ \hline \square\ 5\ 0 \\ \square\square\square\ 0 \\ \hline \square\square\square\ 0 \end{array}$ Answer:	
3. Multiply. $260(400)$ $26 \cdot 4 = \boxed{}$ Attach the 3 ending zeros from the factors to the end of the product. $260(400) = \boxed{},\boxed{}$ Answer:	

Example	Notes
4. Estimate by rounding to the nearest ten. $$7227(87) \approx 7230(\boxed{})$$ $$723 \cdot \boxed{} = \boxed{}$$ Attach the 2 ending zeros from the factors to the end of the product $$7227(87) \approx \boxed{}\boxed{}\boxed{},\boxed{}\boxed{}\boxed{}$$	

5. Divide the following. Set up the division.

$$1482 \div 12$$

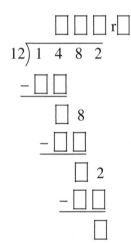

Helpful Hints
When multiplying or dividing whole numbers with several digits, be careful to neatly stack the numbers vertically so that digits in the same place value line up.

Concept Check
1. When multiplying numbers ending in zeros, do you have to write each entire number in columns before multiplying? What can you do to make the multiplication simpler?

Practice
Calculate the following products.

2. $31 \cdot 19$

3. $24 \cdot 225$

Divide the following.

4. $1768 \div 12$

5. $1876 \div 16$

Whole Numbers
Topic 1.9 Exponents

Vocabulary
exponent • base • squared • cubed • power of 1 • power of 0
exponential growth

1. A(n) _____ is a shortcut for repeated multiplication.

2. Any non-zero number raised to the _____ is equal to 1.

Step-by-Step Video Notes
Watch the Step-by-Step Video lesson and complete the examples below.

Example	Notes
1. Identify the base and exponent, then evaluate. 5^2 The base is ☐. The exponent is ☐. $5^2 = 5 \cdot 5 = \boxed{}$ Answer:	
4. Write the following using exponents, then evaluate. $(3)(3)(3)$ $(3)(3)(3) = 3^{\square} = \boxed{}$ Answer:	

Example	Notes
11. Evaluate. 31^0 Answer:	
12. Calculate 10^2, 10^3 and 10^6. $10^2 = \boxed{}$ $10^3 = \boxed{}$ $10^6 = \boxed{}$	

Helpful Hints

The base is the number, or factor, being multiplied by itself. The exponent is the number of times the base is used as a factor.

Any number raised to the power of 1 is the number itself. Any non-zero number raised to the power of 0 is equal to 1.

An exponent expression, for example, 7^2 does NOT mean $7 \cdot 2$. It means use 7 as a factor 2 times, in other words, $7 \cdot 7$. The value of the expression is 49.

Concept Check

1. Can you explain a simple rule for evaluating powers of ten? How does the exponent relate to the number of zeros in the standard form number? Give an example.

Practice

Identify the base and exponent, then evaluate.

2. 4^3

3. 9^2

Evaluate.

4. 10^7

5. 25^0

Whole Numbers
Topic 1.10 Order of Operations and Whole Numbers

Vocabulary
order of operations   •   PEMDAS   •   parentheses   •   exponents

1. When simplifying an expression using the order of operations, always evaluate what is inside _____ or other grouping symbols first.

Step-by-Step Video Notes
Watch the Step-by-Step Video lesson and complete the examples below.

Example	Notes
1. Calculate. $4 + 6 \cdot 3$ The operations in this expression are addition and multiplication. Multiply first, and then add. $4 + \square = \square$ Answer:	
3. Simplify by using the order of operations. $3^2 + 5 \cdot 4$ There is an exponent in the expression and multiplication. Evaluate the exponent then multiply. $\square + 5 \cdot 4 = \square + \square = \square$ Answer:	

Example	Notes

4. Simplify.

$3(4-2)^2 +5$

First simplify the operation in the
_____, and then square that number.

Next _____ by ☐ , then

_____ ☐ .

$3(\boxed{})^2 +5 = 3\cdot\boxed{}+5$

$=\boxed{}+5$

$=\boxed{}$

Answer:

---

6. Simplify.

$\dfrac{22+10}{4\cdot5-4}$

Simplify the numerator. $\dfrac{\boxed{}}{4\cdot5-4}$

Simplify the denominator. $\dfrac{\boxed{}}{\boxed{}}$

Divide. $\boxed{}$

Helpful Hints
These memory tips might help you remember the order of operations. P E MD AS, and Please Excuse My Dear Aunt Sally.

Other symbols also act like parentheses: brackets such as [ ] and { }, and fraction bars. When you see a fraction bar, act like there are parentheses around the numerator and denominator.

Concept Check
1. The order of operations tells you to multiply and divide from left to right. Does the answer change if you multiply first if simplifying an expression such as $6\div3\cdot2$?

Practice
Evaluate.

2. $10-3\cdot2$

3. $4(9-6)^2 -8$

Simplify.

4. $5^2 +9\div3$

5. $\dfrac{14+13}{6\cdot2-3}$

Whole Numbers
Topic 1.11 More Problem Solving

Vocabulary

area   •   perimeter   •   problem solving   •   translating

1.   The _____ of a rectangle is found by multiplying its length times its width.

Step-by-Step Video Notes
Watch the Step-by-Step Video lesson and complete the examples below.

Example	Notes
1. Translate twice the sum of 7 and 4 to symbols. Enter the appropriate operation symbol. $2\left(7\,\square\,4\right)$ Simplify. Answer:	
4. Find the area of a room that has a length of 21 feet and a width of 14 feet. Understand the problem. We need to find the _____ of the room. Create a plan. We will find the _____ by multiplying the length $\square$ feet by the width $\square$ feet. Find the answer. Check. Answer:	

Example	Notes
5. Desmond makes a salary of $39,000 per year. What is his monthly salary? We need to find _____. There are ☐ months in a year. We will _____ his yearly salary by ☐. Answer:	
6. The phone company charges $20 for installation and $25 a month for basic service. How much will Monique pay for one year of service with installation? Answer:	

Helpful Hints

When solving a problem, there are key words or phrases which can be translated into an operation.

After obtaining an answer from your calculation, make sure to check that this answers the question which was asked in the problem.

Concept Check

1. What two steps need to be done before calculation in solving a problem?

Practice

Find the area of the room with the following dimensions.

2. length of 21 feet and width of 12 feet

3. length of 10 feet and width of 8 feet

Joe earns $900 at his summer job where he works for three weeks, 20 hours each week.

4. How much did he earn each week?

5. How much did he earn per hour?

Name: _____   Date: _____
Instructor: _____   Section: _____

Factors and Fractions
Topic 2.1 Factors

Vocabulary
factor   •   divisibility rule(s)   •   common factor   •   greatest common factor (GCF)

1. A _____ is a number that divides exactly into two or more numbers.

Step-by-Step Video Notes
Watch the Step-by-Step Video lesson and complete the examples below.

Example	Notes
3. Find the greatest common factor, or GCF, of 15 and 24. List the factors of 15. 1, ☐, ☐, 15 List the factors of 24. 1, 2, ☐, ☐, ☐, 8, ☐, ☐ Identify the factors common to both lists. 1, ☐ Choose the largest of these. ☐ Answer:	
4. Find the GCF of 18 and 54. List the factors of 18. 1, 2, ☐, ☐, ☐, ☐ List the factors of 54. 1, 2, ☐, ☐, ☐, ☐, ☐, ☐ Identify the factors common to both lists. Choose the largest of these. Answer:	

Example	Notes
5. Find the GCF of 132 and 198.	
6. List three ways to factor 18. List the factors of 18. 1, 2, ☐, ☐, ☐, ☐ Pair two factors that multiply to equal 18. $1 \cdot \square = 18$ Pair two different factors that equal 18. $2 \cdot \square = 18$ Repeat a third time. $\square \cdot \square = 18$	

Helpful Hints

There can be one or more common factors between two or more numbers, but there is only one greatest common factor (GCF).

The word "factor" can be either a noun or verb: Factors are the numbers being multiplied, and to factor a number means to write it as a product.

Concept Check

1. How many common factors are there between 24 and 96? Which is the GCF?

Practice

2. List all factors of 36.

3. Find the GCF of 30 and 85.

4. List three ways to factor 99.

Name: _____   Date: _____
Instructor: _____   Section: _____

Factors and Fractions
Topic 2.2 Prime Factorization

Vocabulary
factor   •   prime number   •   composite number   •   prime factorization

1. A _____ is a whole number greater than 1 that has exactly two factors, the number itself and 1.

Step-by-Step Video Notes
Watch the Step-by-Step Video lesson and complete the examples below.

Example	Notes
1. Find the factors of 36. 1, 2, 3, 4, 6, ☐, ☐, ☐, 36	
3. Find the prime factorization of 24.	

Example	Notes
4. Find the prime factorization of 60.	

Helpful Hints

There is more than one way to write the factorization of a composite number. However, if all factors are rewritten as products of prime numbers, there is only one answer.

In the first step of rewriting a composite number as a product of two factors, there is often more than one combination of two factors that can be selected. However, continuing the steps of prime factorization will lead to the same final answer.

Concept Check

1. In finding the prime factorization of 72, if 72 is written as $6 \cdot 12$, what is the following step? And the next step?

Practice

2. Find the prime factorization of 56.

3. Find the prime factorization of 98.

4. Find the prime factorization of 210.

Factors and Fractions
Topic 2.3 Understanding Fractions

Vocabulary
fraction   •   denominator   •   undefined   •   improper fraction
proper fraction   •   numerator

1. A(n) _____ is the bottom number in a fraction and indicates the number of parts in the whole.

2. A(n) _____ is a fraction where the numerator is greater than or equal to the denominator.

Step-by-Step Video Notes
Watch the Step-by-Step Video lesson and complete the examples below.

Example	Notes
2a. How many parts are shaded in the diagram?	
How many total parts are in the whole diagram? ☐	
Write a fraction that represents the shaded part.	
Answer:	
2b. Write a fraction that represents the shaded part.	

Example	Notes
3. The numerator of $\frac{4}{9}$ is ___ the denominator. Identify this fraction as proper or improper. Answer:	
6. Identify $\frac{3}{3}$ as proper or improper. Answer:	

Helpful Hints
When the numerator is equal to the denominator of a fraction, this is an improper fraction that is equal to 1.

A fraction with a numerator of zero is equal to zero; a fraction with a denominator of zero is undefined.

Concept Check
1. There are 9 boys in the class of 20 students. Represent this as a fraction and state why this is a proper fraction.

Practice
2. Write a fraction that represents the shaded part.

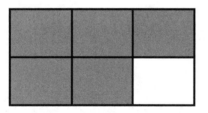

3. Identify $\frac{7}{7}$ as proper or improper.

4. Identify $\frac{2}{9}$ as proper or improper.

5. Identify $\frac{0}{5}$ as proper or improper.

Factors and Fractions
Topic 2.4 Simplifying Fractions – GCF and Factors Method

Vocabulary
equivalent fractions • factors • simplest form of a fraction • greatest common factor

1. _____ are fractions that represent the same value.

2. A fraction is said to be in _____ if there is no common factor other than 1 that divides exactly into the numerator and the denominator.

Step-by-Step Video Notes
Watch the Step-by-Step Video lesson and complete the examples below.

Example	Notes
1. Write $\dfrac{9}{15}$ in simplest form using the GCF method. Find the GCF of 9 and 15. ☐ Divide the numerator and the denominator by the GCF. Answer:	
2. Write $\dfrac{72}{96}$ in simplest form using the GCF method.	

Example	Notes
3. Write $\dfrac{20}{32}$ in simplest form using the factors method. Name a common factor of 20 and 32. $\boxed{2}$ Divide numerator and denominator by 2. $\dfrac{\boxed{}}{16}$ Name a common factor of $\boxed{}$ and 16. $\boxed{}$ Divide the numerator and denominator by this common factor. Answer:	

Helpful Hints

Dividing the numerator and denominator by a common factor results in an equivalent fraction; however, this is not always the simplest form of the fraction. The procedure must be repeated until there is no common factor of the numerator and denominator except for 1.

Using the GCF method leads to a quicker solution process than the factors method; however, if you are having a hard time finding the GCF, then the factors method is helpful.

Concept Check

1. Which method would you use for simplifying $\dfrac{6}{9}$? For simplifying $\dfrac{48}{175}$? Describe the first step for each method selected for the above fraction simplifications.

Practice

2. Write $\dfrac{24}{27}$ in simplest form using the GCF method.

3. Write $\dfrac{35}{175}$ in simplest form using the factors method.

4. Write $\dfrac{96}{144}$ in simplest form using either the GCF method or the factors method.

Factors and Fractions
Topic 2.5 Simplifying Fractions – Prime Factors Method

Vocabulary
prime factors method   •   composite number   •   common factor   •   simplest form

1. The prime factor method for simplifying fractions involves writing the numerator and denominator as products of prime numbers, then dividing the numerator and denominator by the _____.

Step-by-Step Video Notes
Watch the Step-by-Step Video lesson and complete the examples below.

Example	Notes
1. Write $\dfrac{24}{36}$ in simplest form using the prime factors method. Write the numerator as the product of prime numbers. $24 = 2^3 \cdot \square$ Write the denominator as the product of prime numbers. $36 = 2^2 \cdot \square$ Divide the numerator and the denominator by the common factors. Answer:	

Example	Notes
2. Write in simplest form using the prime factors method. $\dfrac{84}{91}$ Write the numerator as a product of prime numbers. ☐ Write the denominator as a product of prime numbers. ☐ Divide the numerator and the denominator by the GCF. Multiply the remaining factors. Answer:	
3. Write $\dfrac{39}{135}$ in simplest form using the prime factors method.	

Helpful Hints

When simplifying fractions using the prime factor method, you must make sure that both the numerator and the denominator are written as the product of only prime numbers.

If the numerator is a factor of the denominator, then the simplified fraction will have numerator of 1.

Concept Check

1. How is the prime factors method for simplifying fractions similar to the GCF method?

Practice

2. Write $\dfrac{24}{27}$ in simplest form using the prime factors method.

3. Write $\dfrac{35}{175}$ in simplest form using the prime factors method.

4. Write $\dfrac{20}{120}$ in simplest form using the prime factors method.

Factors and Fractions
Topic 2.6 Multiplying Fractions

Vocabulary
numerator   •   denominator   •   multiplying fractions   •   equivalent fraction

1. Multiplying fractions can be done by first multiplying the numerators to get the
    _____ of the product and multiplying the denominators to get the
    _____ of the product.

Step-by-Step Video Notes
Watch the Step-by-Step Video lesson and complete the examples below.

Example	Notes
1. Multiply. Write your answer in simplest form. $\dfrac{4}{7}\cdot\dfrac{3}{5}$ Multiply the numerators. $4\cdot 3=\square$ Multiply the denominators. $7\cdot\square=\square$ Simplify. Answer:	
2. Multiply. Write your answer in simplest form. $\dfrac{7}{10}\cdot\dfrac{5}{8}$ Answer:	

Example	Notes
4. Multiply $\dfrac{4}{5} \cdot \dfrac{9}{16}$ by simplifying first. Write as one fraction. Do not multiply yet. Divide common factors in the numerator and denominator. Multiply the remaining factors. Answer:	
5. Multiply $3 \cdot \dfrac{2}{9}$ by simplifying first.	

Helpful Hints

Both the numerators and the denominators must be multiplied when multiplying fractions; fractions should be written in simplest form.

When fractions are being multiplied and there are common factors in one or more of the numerators with one or more of the denominators, it is easier to simplify the fractions by dividing the numerator and the denominator by the common factors before multiplying.

Concept Check

1. Why is $\dfrac{7}{100} \cdot \dfrac{10}{21}$ an easier multiplication problem than $\dfrac{5}{16} \cdot \dfrac{7}{11}$?

Practice

2. Multiply $\dfrac{8}{15} \cdot \dfrac{3}{5}$. Write your answer in simplest form.

3. Multiply $\dfrac{2}{9} \cdot \dfrac{3}{5}$ by simplifying first.

4. Multiply $7 \cdot \dfrac{3}{28}$ by simplifying first.

Factors and Fractions
Topic 2.7 Dividing Fractions

Vocabulary
reciprocals   •   common factors

1. Two numbers are _____ of each other if their product is 1.

Step-by-Step Video Notes
Watch the Step-by-Step Video lesson and complete the examples below.

Example	Notes
1. Find the reciprocal of $\frac{3}{4}$. The reciprocal is $\frac{4}{\square}$	
2. Find the reciprocal of 5.	
4. Divide $\frac{4}{5} \div \frac{3}{8}$. Write in simplest form. Take the reciprocal of the second fraction. $\frac{\square}{\square}$ Multiply the reciprocal by the first fraction. $\frac{4}{5} \cdot \frac{\square}{\square} = \frac{\square}{\square}$ Answer: _____	

Example	Notes
6. Divide $12 \div \dfrac{2}{3}$. Write in simplest form. Answer:	

Helpful Hints

To find the reciprocal of a whole number, first write the number as a fraction by putting it over 1.

When fractions are being divided, the reciprocal of the second fraction must be taken before the fractions are multiplied.

The numerator and denominator can be divided by common factors when dividing fractions, but only after the division has been changed to multiplication by the inverted second fraction.

Concept Check

1. How can $\dfrac{2}{5} \div \dfrac{3}{7}$ be rewritten as a multiplication?

Practice

2. Divide $\dfrac{8}{15} \div \dfrac{3}{4}$. Write your answer in simplest form.

3. Divide $8 \div \dfrac{2}{5}$. Write your answer in simplest form.

4. Divide $\dfrac{2}{7} \div 6$. Write your answer in simplest form.

Name: _____   Date: _____

Instructor: _____   Section: _____

LCM and Fractions
Topic 3.1 Finding the LCM – List Method

Vocabulary
multiple   •   common multiple   •   least common multiple   •   whole number

1. A _____ of a number is the product of a number and a positive whole number.

2. The _____ of two or more numbers is the smallest number that is a multiple of the given numbers.

Step-by-Step Video Notes
Watch the Step-by-Step Video lesson and complete the examples below.

Example	Notes
1. Find the first five multiples of 9. 9, 18, ☐, 36, 45	
2. Find the first four multiples of 20. 20, ☐, ☐, ☐	
3. Find the least common multiple of 9 and 12. List the first several multiples of 9. 9, 18, ☐, 36, 45, ☐, ☐, 72 List the first several multiples of 12. 12, 24, ☐, ☐, 60, 72 Identify the multiples common to each list. ☐, 72 Choose the least of these. Answer:	

Example	Notes
4. Find the least common multiple of 6, 8 and 12.	

Helpful Hints

We can find a common multiple of two numbers by multiplying them together, but this may or may not be the least common multiple.

Be careful not to confuse the GCF, the greatest common factor, with the LCM, the least common multiple; the GCF is the greatest factor that can be divided evenly into the given numbers, while the LCM is the smallest number that is a multiple of the given numbers.

The GCF of two numbers is always less than or equal to the given numbers, while the LCM is always greater than or equal to the given numbers.

Concept Check

1. The following statement is false: The LCM of 4 and 10 is 40 because $4 \cdot 10 = 40$. Why is this a false statement?

Practice

2. List the first five multiples of 16.

3. List the first five multiples of 20.

4. Find the least common multiple of 16 and 20.

5. Find the least common multiple of 7, 9, and 21.

LCM and Fractions
Topic 3.2 Finding the LCM – GCF Method

Vocabulary
multiply   •   greatest common factor   •   least common multiple   •   divide

1. When finding the LCM of two numbers using the GCF method, first find the GCF of the two numbers, then _____ the two original numbers and divide by the GCF.

Step-by-Step Video Notes
Watch the Step-by-Step Video lesson and complete the examples below.

Example	Notes
1. Find the LCM of 6 and 9 using the GCF method. Find the GCF of 6 and 9. ☐ Multiply 6 and 9. $6 \cdot 9 = $ ☐ Divide product by GCF. Answer:	
2. Find the LCM of 6 and 15 using the GCF method. Find the GCF of 6 and 15. ☐ Multiply 6 and 15. ☐ $\cdot$ ☐ $=$ ☐ Divide product by GCF. Answer:	

Example	Notes
3. Find the LCM of 4 and 18 using the GCF method.	
4. Find the LCM of 7 and 9 using the GCF method.	

Helpful Hints
If the GCF of two numbers is 1, then the LCM is the product of the two numbers.

Finding the LCM of two or more numbers using the GCF method can be quicker than listing out multiples of each number.

Concept Check
1. Listing multiples of 3 and 42 to determine the LCM leads to a long list for the number 3. List the steps needed for finding these two numbers using the GCF method.

Practice
2. Find the LCM of 6 and 42 using the GCF method.

3. Find the LCM of 10 and 25 using the GCF method.

4. Find the LCM of 8 and 20 using the GCF method.

5. Find the LCM of 5 and 11 using the GCF method.

LCM and Fractions
Topic 3.3 Finding the LCM – Prime Factor Method

Vocabulary
prime factor method   •   common prime factors   •   prime factorization   •   multiples

1. The _____ of any whole number is the factored form in which all factors are prime numbers.

Step-by-Step Video Notes
Watch the Step-by-Step Video lesson and complete the examples below.

Example	Notes
3. Find the LCM of 12 and 42 using the prime factor method. Find the prime factorization of 12. $12 = 2 \cdot 2 \cdot 3$ Find the prime factorization of 42. $42 = 2 \cdot 3 \cdot \underline{\quad}$ Write the prime factorizations one below the other, putting the common prime factors below each other.	

2	2	3	
2		3	☐

Write down the prime factor from each column.

2	2	3	☐

Multiply the list of prime factors.

Answer:

Example	Notes
4. Find the LCM of 8 and 30 using the prime factor method. Write the prime factorizations of 8 and 30 one below the other, putting the common prime factors below each other. Multiply the list of prime factors. Answer:	
6. Find the LCM of 13 and 19 using the prime factor method.	

Helpful Hints

When using the prime factor method to find the LCM of two numbers, make sure to enter a prime factor in a column for each time it is used. Do not enter a prime factor with an exponent.

When using the prime factor method to find the LCM of two numbers, make sure to enter a prime factor from each column, even if the prime factor is not a factor of both of the original numbers.

Concept Check
1. How many total prime factors (matching the total number of columns) are there when finding the LCM of 24 and 30?

Practice
2. Find the LCM of 3 and 42 using the prime factor method.

3. Find the LCM of 10 and 25 using the prime factor method.

4. Find the LCM of 15 and 14 using the prime factor method.

5. Find the LCM of 8 and 9 using the prime factor method.

LCM and Fractions
Topic 3.4 Writing Fractions with an LCD

Vocabulary
least common denominator (LCD) • prime numbers • common factor • equivalent fractions

1. The _____ is the least common multiple (LCM) of the denominators.

2. Fractions are _____ if they have the same value.

Step-by-Step Video Notes
Watch the Step-by-Step Video lesson and complete the examples below.

Example	Notes
1. Find the LCM of 4 and 12. ☐ Find the LCD of $\dfrac{3}{4}$ and $\dfrac{5}{12}$. Answer:	
3. Write a fraction equivalent to $\dfrac{1}{6}$ using denominator of 12. What number does 6 need to be multiplied by to result in 12? ☐ Multiply the numerator and denominator of the original fraction by the same number. $\dfrac{1}{6} \cdot \dfrac{\square}{\square} = \dfrac{\square}{12}$ Answer:	

Example	Notes

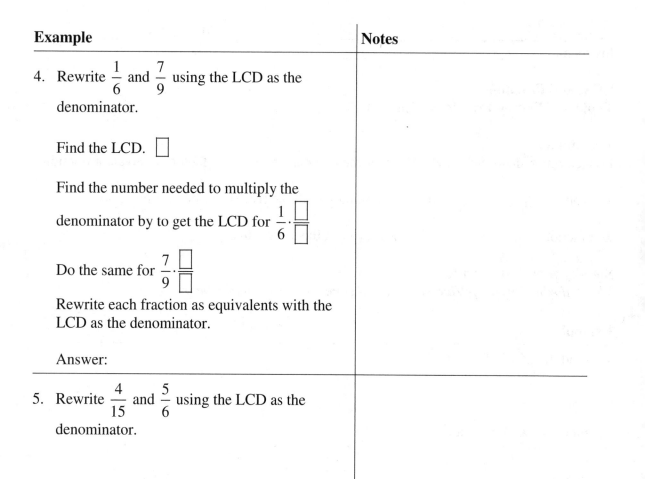

4. Rewrite $\dfrac{1}{6}$ and $\dfrac{7}{9}$ using the LCD as the denominator.

Find the LCD. ☐

Find the number needed to multiply the denominator by to get the LCD for $\dfrac{1}{6} \cdot \dfrac{\boxed{}}{\boxed{}}$

Do the same for $\dfrac{7}{9} \cdot \dfrac{\boxed{}}{\boxed{}}$

Rewrite each fraction as equivalents with the LCD as the denominator.

Answer:

5. Rewrite $\dfrac{4}{15}$ and $\dfrac{5}{6}$ using the LCD as the denominator.

Helpful Hints

If one denominator divides exactly into the other, then the LCD is the larger number.

If two denominators have no common factor other than 1, then the LCD is the product of the two denominators.

Concept Check

1. When rewriting $\dfrac{1}{3}$ and $\dfrac{1}{5}$ using the LCD, how do you know that neither numerator will be 1?

Practice

2. Rewrite $\dfrac{2}{3}$ using 9 as the denominator.

3. Rewrite $\dfrac{5}{9}$ and $\dfrac{2}{15}$ using the LCD as the denominator.

4. Rewrite $\dfrac{1}{4}$ and $\dfrac{5}{6}$ using the LCD as the denominator.

Name: _____   Date: _____

Instructor: _____   Section: _____

LCM and Fractions
Topic 3.5 Adding and Subtracting Like Fractions

Vocabulary
like fractions   •   unlike fractions   •   numerators   •   equivalent fractions

1. Fractions with the same, or common, denominator are called _____.

2. Fractions without a common denominator are called _____.

Step-by-Step Video Notes
Watch the Step-by-Step Video lesson and complete the examples below.

Example	Notes
3. Add the like fractions $\frac{2}{5} + \frac{1}{5}$. Simplify if possible. Add the numerators. $2 + 1 = \square$ Keep the denominator. $\square$ Answer:	
4. Add the like fractions $\frac{3}{10} + \frac{3}{10}$. Simplify if possible.	

45

Example	Notes
5. Subtract the like fractions $\frac{6}{7} - \frac{2}{7}$. Simplify if possible. Subtract the numerators. $6 - 2 = \square$ Keep the denominator. $\square$ Answer:	
6. Subtract the like fractions $\frac{11}{14} - \frac{5}{14}$. Simplify if possible.	

Helpful Hints

When adding or subtracting like fractions, the operation is done on the numerators only.

When adding or subtracting like fractions, the denominator remains unchanged.

Concept Check

1. Whether adding or subtracting $\frac{4}{5}$ and $\frac{2}{5}$, what is the denominator of the resulting fraction? What will the numerator of the result of addition be? What will the numerator of the result of subtraction be?

Practice

2. Add the like fractions $\frac{4}{11} + \frac{3}{11}$. Simplify if possible.

3. Add the like fractions $\frac{1}{18} + \frac{2}{18} + \frac{11}{18}$. Simplify if possible.

4. Subtract the like fractions $\frac{5}{9} - \frac{1}{9}$. Simplify if possible.

5. Subtract the like fractions $\frac{5}{8} - \frac{3}{8}$. Simplify if possible.

LCM and Fractions
Topic 3.6 Adding and Subtracting Unlike Fractions

Vocabulary
like fractions   •   unlike fractions   •   prime factors   •   denominator

1. Fractions without a common denominator are called _____.

Step-by-Step Video Notes
Watch the Step-by-Step Video lesson and complete the examples below.

Example	**Notes**
1. Add $\frac{3}{4}+\frac{1}{6}$. Simplify if possible. Find the LCD. 12 Rewrite as like fractions with the LCD of 12 as the denominator. $\frac{3}{4}=\frac{\square}{12}$, $\frac{1}{6}=\frac{\square}{12}$ Add the numerators. $\square$ Keep the denominator. $\square$ Answer:	
2. Add $\frac{3}{8}+\frac{2}{5}$. Simplify if possible.	

Example	Notes
3. Subtract $\dfrac{5}{6} - \dfrac{1}{3}$. Simplify if possible. Find the LCD. ☐ Rewrite the fractions as like fractions with the LCD as the denominator. Subtract the numerators. Keep the denominator. Answer:	
4. Subtract $\dfrac{7}{10} - \dfrac{4}{15}$. Simplify if possible.	

Helpful Hints

When rewriting fractions with a common denominator, the LCD is usually used.

When adding or subtracting unlike fractions, a common denominator is needed.

Concept Check

What important step(s) must be done before adding or subtracting unlike fractions?

Practice

1. Add $\dfrac{2}{3} + \dfrac{1}{5}$. Simplify if possible.

2. Add $\dfrac{5}{6} + \dfrac{1}{18}$. Simplify if possible.

3. Subtract $\dfrac{3}{4} - \dfrac{1}{10}$. Simplify if possible.

4. Subtract $\dfrac{5}{7} - \dfrac{3}{14}$. Simplify if possible.

LCM and Fractions
Topic 3.7 Order of Operations and Fractions

Vocabulary
order of operations   •   addition of fractions   •   multiplication of fractions   •   LCM

1. The procedure for the _____ is to evaluate what is inside parentheses, evaluate any exponents, perform multiplication/division, and then perform addition/subtraction.

Step-by-Step Video Notes
Watch the Step-by-Step Video lesson and complete the examples below.

Example	Notes
1. Simplify $\left(\dfrac{1}{3}-\dfrac{1}{6}\right)+\dfrac{1}{2}$ by using the order of operations. Evaluate what is inside the parentheses. $\dfrac{1}{3}-\dfrac{1}{6}=\dfrac{\Box}{\Box}$ Now perform the addition. $\dfrac{\Box}{\Box}+\dfrac{1}{2}=\dfrac{\Box}{\Box}$ Answer:	
3. Simplify $\dfrac{1}{4}\div\dfrac{3}{2}+\dfrac{1}{2}\cdot\dfrac{1}{3}$ by using the order of operations. Perform the multiplication and division from left to right. $\dfrac{1}{4}\div\dfrac{3}{2}=\dfrac{\Box}{\Box}$, $\dfrac{1}{2}\cdot\dfrac{1}{3}=\dfrac{1}{6}$ Now perform the addition. $\dfrac{\Box}{\Box}+\dfrac{1}{6}=\dfrac{\Box}{\Box}$ Answer:	

Example	Notes
4. Simplify $\left(\dfrac{3}{4}\right)\left(\dfrac{1}{2}-\dfrac{1}{4}\right)^2+\dfrac{2}{5}\cdot\dfrac{1}{2}$ using the order of operations.	

Helpful Hints

PE MD AS is an acronym to help remember the order of operations. This can be remembered as Please Excuse My Dear Aunt Sally; the letter P stands for parentheses, E for exponents, MD for multiplication/division, and AS for addition/subtraction.

After addressing any parentheses and exponents, make sure to do the multiplication/division from left to right, then the addition/subtraction from left to right.

Concept Check

1. Why is the first step to simplifying $\dfrac{1}{5}+\dfrac{2}{5}\cdot\left(\dfrac{1}{4}\right)^2$ not adding $\dfrac{1}{5}+\dfrac{2}{5}$, nor multiplying $\dfrac{2}{5}\cdot\dfrac{1}{4}$?
   What is the second step? The third?

Practice

2. Simplify $\dfrac{1}{7}+\dfrac{2}{7}\cdot\left(\dfrac{1}{2}\right)^2$ using the order of operations.

3. Simplify $\dfrac{6}{7}-\dfrac{1}{3}\div\dfrac{5}{12}$ using the order of operations.

4. Simplify $\dfrac{1}{4}\cdot\left(\dfrac{3}{5}+\dfrac{1}{10}\right)\div\dfrac{3}{10}$ using the order of operations.

Mixed Numbers
Topic 4.1 Changing a Mixed Number to an Improper Fraction

Vocabulary
mixed number   •   improper fraction   •   numerator   •   whole number

1. A(n) _____ is the sum of a whole number and a fraction.

Step-by-Step Video Notes
Watch the Step-by-Step Video lesson and complete the examples below.

Example	Notes
1. Write $2\frac{1}{3}$ as an improper fraction. Multiply the denominator by the whole number. $\square \times \square = 6$ Add this to the numerator. $6 + \square = \square$ Write this value over the original denominator. $\dfrac{\square}{3}$ Answer:	
2. Write $4\frac{5}{7}$ as an improper fraction. Multiply the denominator by the whole numerator. Add this to the numerator. Answer:	

Example	Notes
4. Write 25 as an improper fraction.	
Answer:	
6. Write 91 as an improper fraction. Choose a denominator other than 1.	
Answer:	

Helpful Hints

When changing a mixed number to an improper fraction, the denominator will always be the same.

Whole numbers can be changed to improper fractions having any denominator desired; the numerator is multiplied by the chosen denominator.

Concept Check

1. When changing $2\frac{4}{5}$ to an improper fraction, explain why the numerator will be 14.

Practice

Write as an improper fraction.

2. $5\frac{9}{20}$

3. $10\frac{4}{7}$

4. 82

5. $12\frac{1}{6}$

Mixed Numbers
Topic 4.2 Changing an Improper Fraction to a Mixed Number

Vocabulary
remainder   •   quotient   •   numerator   •   denominator

1. When changing an improper fraction to a mixed number, the _____ is the whole number part.

Step-by-Step Video Notes
Watch the Step-by-Step Video lesson and complete the examples below.

Example	Notes
1. Write $\dfrac{13}{5}$ as a division problem. $13 \div \square$ Answer:	
2. Write $\dfrac{13}{4}$ as a mixed number. Divide the numerator by the denominator. $\dfrac{\square}{4\overline{)13}}$ What is the remainder? $\square$ The quotient is the whole number part. The remainder is the numerator of the fractional part. The denominator stays the same. $\square\dfrac{\square}{4}$ Answer:	

Example	Notes
3. Write $\dfrac{7}{5}$ as a mixed number. Answer:	
4. Write $\dfrac{19}{5}$ as a mixed number. Answer:	

Helpful Hints

When setting up a division problem to change an improper fraction into a mixed number, the top number (the numerator) goes inside the long division sign.

If the remainder is zero when changing an improper fraction into a mixed number, then the answer is a whole number; if there is a remainder, this becomes the numerator of the mixed number.

Concept Check

1. When changing an improper fraction to a mixed number, what does the whole number represent?

Practice

Write each as a mixed number or whole number.

2. $\dfrac{11}{3}$

3. $\dfrac{17}{4}$

4. $\dfrac{24}{6}$

5. $\dfrac{25}{7}$

Name: _____   Date: _____

Instructor: _____   Section: _____

Mixed Numbers
Topic 4.3 Multiplying Mixed Numbers

Vocabulary
factor   •   mixed number   •   whole number   •   improper fraction

1. When multiplying mixed numbers, the first step is to rewrite each mixed number as a(n) _____.

Step-by-Step Video Notes
Watch the Step-by-Step Video lesson and complete the examples below.

Example	Notes
1. Multiply $1\frac{1}{2} \cdot 5\frac{1}{6}$. Write $1\frac{1}{2}$ and $5\frac{1}{6}$ as improper fractions. $1\frac{1}{2} = \dfrac{\square}{2}$ $5\frac{1}{6} = \dfrac{\square}{6}$ Multiply the fractions. $\dfrac{\square}{2} \cdot \dfrac{\square}{\square} = \dfrac{\square}{\square}$ Rewrite as a mixed or whole number. Answer:	
2. Multiply $1\frac{2}{3} \cdot 4\frac{1}{5}$. Write each mixed number as an improper fraction. $\dfrac{\square}{3} \cdot \dfrac{\square}{5}$ Multiply and rewrite answer as a mixed or whole number. Answer:	

Example	Notes
4. Multiply $2\left(3\dfrac{1}{2}\right)\left(1\dfrac{2}{7}\right)$. Answer:	

Helpful Hints

Remember when multiplying fractions, the numerators are multiplied and the denominators are multiplied.

Make sure to simplify the fraction answer in either the resulting improper fraction or the fraction of the mixed number.

Concept Check

1. When multiplying mixed numbers, what form of the numbers must be used?

Practice

Multiply.

2. $2\dfrac{2}{3} \cdot 1\dfrac{1}{4}$

4. $1\dfrac{5}{6} \cdot 2\dfrac{4}{7}$

3. $1\dfrac{2}{7} \cdot 2\dfrac{1}{3}$

5. $3\left(1\dfrac{2}{3}\right)\left(2\dfrac{4}{5}\right)$

Mixed Numbers
Topic 4.4 Dividing Mixed Numbers

Vocabulary
mixed number   •   whole number   •   factor   •   improper fraction

1. When dividing mixed numbers, each mixed number should be written as a(n)
    _____.

Step-by-Step Video Notes
Watch the Step-by-Step Video lesson and complete the examples below.

Example	Notes
1. Divide. $1\frac{3}{4} \div 2\frac{4}{5}$	

Rewrite $1\frac{3}{4}$ as an improper fraction. $\dfrac{\square}{4}$

Rewrite $2\frac{4}{5}$ as an improper fraction. $\dfrac{\square}{5}$

Divide the fractions by inverting the second fraction and multiplying it by the first fraction.

$$\frac{\square}{4} \cdot \frac{\square}{\square} = \frac{\square}{\square}$$

Rewrite answer as a mixed or whole number.

Answer:

Example	Notes
2. Divide. $3\dfrac{2}{5} \div 2\dfrac{2}{3}$ Write each mixed number as an improper fraction. $\dfrac{\square}{5} \div \dfrac{\square}{3}$ Divide the fractions. Rewrite answer as a mixed or whole number. Answer:	
4. Divide. $3\dfrac{2}{5} \div \dfrac{1}{5}$ Answer:	

Helpful Hints

Remember that whole numbers can be rewritten as improper fractions by placing the original number as the numerator and making the denominator 1.

Always simplify a fraction by dividing both numerator and denominator by common factors.

Concept Check
1. Why must mixed numbers be changed to improper fractions before dividing them?

Practice
Divide.

2. $1\dfrac{1}{4} \div 3\dfrac{1}{3}$

4. $6 \div 2\dfrac{4}{5}$

3. $4\dfrac{2}{5} \div 2\dfrac{3}{4}$

5. $3\dfrac{5}{6} \div \dfrac{1}{6}$

Mixed Numbers
Topic 4.5 Adding Mixed Numbers

Vocabulary
mixed number　•　common denominator　•　numerator　•　denominator

1. In order to add fractions, they must have the same _____.

Step-by-Step Video Notes
Watch the Step-by-Step Video lesson and complete the examples below.

Example	Notes
1. Add. $3\frac{1}{3}+2\frac{1}{3}$ Add the like fractions by adding the numerators and keeping the denominator. $\frac{1}{3}+\frac{1}{3}=\frac{\Box}{3}$ Add the whole numbers. $3+2=\Box$ Answer:	
2. Add. $7\frac{1}{3}+4\frac{3}{5}$ Rewrite the fractions with a common denominator. $\frac{1}{3}=\frac{\Box}{\Box}, \qquad \frac{3}{5}=\frac{\Box}{\Box}$ Add the fractions. Add the whole numbers. Answer:	

Example	Notes
4. Add. $2\frac{1}{3} + 15\frac{2}{3}$ When adding fractions yields an improper fraction, it needs to be changed to a mixed or whole number. This is then added to the sum of the whole numbers. $\frac{1}{3} + \frac{2}{3} = \frac{\square}{3} = \square$ Answer:	
6. Add. $2\frac{1}{6} + \frac{3}{4}$ Answer:	

Helpful Hints

Fractions must have the same denominator to be added; fractions can be rewritten with a common denominator, the LCD.

Always simplify fractions.

Concept Check

1. What are the steps used when adding mixed numbers?

Practice

Add.

2. $4\frac{1}{5} + 2\frac{2}{5}$

4. $1\frac{7}{10} + 2\frac{4}{5}$

3. $2\frac{3}{4} + 3\frac{1}{4}$

5. $5\frac{1}{2} + 3\frac{1}{3}$

Name: _____   Date: _____

Instructor: _____   Section: _____

Mixed Numbers
Topic 4.6 Subtracting Mixed Numbers

Vocabulary
mixed number   •   denominator   •   numerator   •   improper fraction

1. In order to subtract fractions, they must have the same _____.

Step-by-Step Video Notes
Watch the Step-by-Step Video lesson and complete the examples below.

Example	Notes
1. Subtract $5\frac{4}{7} - 2\frac{1}{7}$. Subtract the fractions. $\frac{4}{7} - \frac{1}{7} = \frac{\square}{7}$ Subtract the whole numbers. $5 - 2 = \square$ Answer:	
2. Subtract $6\frac{1}{2} - 3\frac{1}{4}$. Rewrite the fractions with the LCD. $\frac{1}{2} = \frac{\square}{\square}$, $\frac{1}{4} = \frac{\square}{\square}$ Subtract the fractions. Subtract the whole numbers. Answer:	

Example	Notes
3. Subtract $9\dfrac{1}{3} - 2\dfrac{5}{8}$.	
Answer:	
5. Subtract $4 - 1\dfrac{1}{2}$.	
Answer:	

Helpful Hints

Borrowing can be needed when subtracting mixed numbers, similar to subtracting integers.

Borrowing one can be rewritten as an improper fraction with the needed denominator, by writing the same number as the numerator.

Concept Check
1. Why must fractions have the same denominator to subtract them?

Practice
Subtract.

2. $2\dfrac{4}{5} - 1\dfrac{2}{5}$

4. $5\dfrac{1}{3} - 2\dfrac{1}{6}$

3. $5\dfrac{1}{3} - 2\dfrac{2}{3}$

5. $6\dfrac{1}{5} - 1\dfrac{3}{10}$

Mixed Numbers
Topic 4.7 Adding and Subtracting Mixed Numbers - Improper Fractions

Vocabulary
mixed numbers   •   improper fractions   •   numerators   •   lowest common denominators

1. When adding and subtracting mixed numbers, the mixed numbers can be rewritten as
    _____.

Step-by-Step Video Notes
Watch the Step-by-Step Video lesson and complete the examples below.

Example	Notes
2. Subtract $3\frac{4}{5} - 2\frac{2}{5}$. Write each mixed number as an improper fraction. $3\frac{4}{5} = \frac{\square}{5}$, $2\frac{2}{5} = \frac{\square}{5}$ Subtract the fractions. $\frac{\square}{5} - \frac{\square}{5} = \frac{\square}{5}$ Change the improper fraction to a mixed number. Answer:	
3. Add $4\frac{2}{3} + 3\frac{2}{3}$. Write each mixed number as an improper fraction and add. $\frac{\square}{3} + \frac{\square}{3} = \frac{\square}{\square}$ Answer:	

Example	Notes

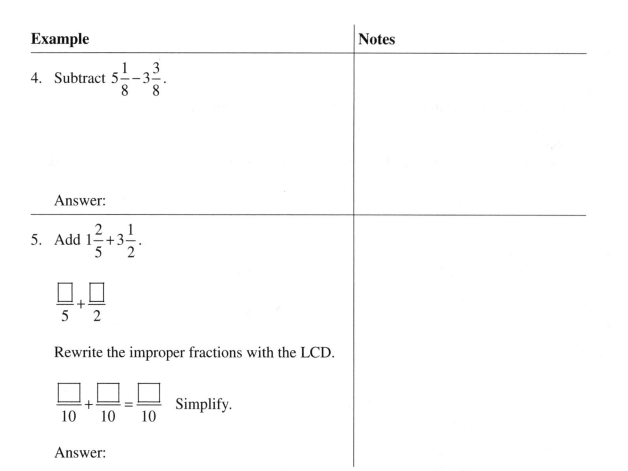

4. Subtract $5\frac{1}{8} - 3\frac{3}{8}$.

Answer:

5. Add $1\frac{2}{5} + 3\frac{1}{2}$.

$$\frac{\square}{5} + \frac{\square}{2}$$

Rewrite the improper fractions with the LCD.

$$\frac{\square}{10} + \frac{\square}{10} = \frac{\square}{10}$$ Simplify.

Answer:

Helpful Hints

Writing mixed numbers as improper fractions before adding, allows you to avoid changing improper fractions to mixed numbers and carrying a one to the whole number part.

Writing mixed numbers as improper fractions before subtracting, allows you to avoid borrowing from 1 from a whole number and convert it to an improper fraction.

Concept Check

1. What is the first step to adding or subtracting mixed numbers?

Practice

Add.

2. $2\frac{5}{7} + 1\frac{3}{7}$

3. $3\frac{1}{4} + 1\frac{1}{2}$

Subtract.

4. $4\frac{2}{3} - 1\frac{1}{3}$

5. $4\frac{2}{3} - 3\frac{1}{6}$

Operations with Decimals
Topic 5.1 Decimal Notation

Vocabulary
place value   •   standard form   •   decimal point   •   decimal fraction

1. A decimal, also known as a decimal number, has three parts: a whole number part, a
    _____ _____, and a decimal part.

Step-by-Step Video Notes
Watch the Step-by-Step Video lesson and complete the examples below.

Example	**Notes**
2. Write the decimal fraction and the decimal number that represents each shaded part.	

The decimal fraction for the shaded part is $\dfrac{\square}{\square}$.

The decimal number for the shaded part is ☐ .

3. Read the decimal 2.4 .

Whole Numbers				.	Decimals				
One Thousands	Hundreds	Tens	Ones	Decimal Point	Tenths	Hundredths	One – Thousandths	Ten – Thousandths	Hundred – Thousandths

Read 2.4 as "two and four _____."

Example	Notes
7. Write the decimal in words. 0.32 0.32 in words is "thirty-two _____."	
11. Write the decimal in standard form. twenty-one and two hundred thirty-seven thousandths 21.☐	

Helpful Hints

When reading or writing a decimal number, use the decimal point in place of the word "and."

When writing decimals in standard form, use the given place value to determine the number of decimal places. Write the decimal part in number form so that it ends at the given place value, inserting zeros at the beginning if needed.

Concept Check

1. How many zeros should you insert after the decimal point when writing the number four and three thousandths in standard form?

Practice

Write each decimal in words.

2. 301.03

Write each decimal in standard form.

4. twenty-seven thousandths

3. 4.718

5. five hundred ten and nine tenths

Name: _____   Date: _____

Instructor: _____   Section: _____

Operations with Decimals
Topic 5.2 Comparing Decimals

Vocabulary
inequality symbols   •   is less than   •   is greater than   •   comparing decimals

1. The symbol $<$ means _____.

2. _____ always point to the smaller number.

Step-by-Step Video Notes
Watch the Step-by-Step Video lesson and complete the examples below.

Example	Notes
1. Which is larger, 0.14 or 0.41? The whole number part is the same for both. Compare the decimal parts, which have the same place value. $41 > 14$, so ☐ $>$ ☐	
3. Which is larger, 0.6 or 0.59? The decimal parts have different place values. Write all decimals with the same number of decimal places, adding zeros to the ends of decimals as needed. $.60 > .59$, so ☐ $>$ ☐	

Example	Notes
5. Fill in the blank with > or < to make a true statement. 4.13 ☐ 4.9	

8. Order the scores from smallest to largest.
 	
    During the 2008 Olympic Games, the U.S. Women Gymnastics Team scored 46.875 on the vault, 47.975 on the uneven bars, 47.25 on the balance beam, and 44.425 on the floor exercises.
 	
    Answer:	

Helpful Hints

Adding zeros to the end of a decimal *does not* change its value. Inserting zeros at the beginning of the decimal part of a number *does* change its value. For example, 0.5 is the same as 0.50, but it is not the same as 0.05.

Concept Check

1. What is a way to compare decimal numbers with a different number of decimal places?

Practice

For each pair of decimals, which is larger?

2. 0.53 or 0.503

Arrange from smallest to largest.

4. 1.104, 1.04, 1.4, 1.14

3. 6.3 or 6.29

5. 0.3, 0.33, 0.303, 0.033

Operations with Decimals
Topic 5.3 Rounding Decimals

Vocabulary
sales tax   •   batting average   •   round up   •   round down

1. You underline the digit in the place value to which you are rounding. If the digit to the right is 5 or more, you will _____, or increase the underlined digit by 1.

Step-by-Step Video Notes
Watch the Step-by-Step Video lesson and complete the examples below.

Example	Notes
1. Rasheed buys a camera for $328.28. There is a sales tax, which calculates to $20.5175. The store needs to round this tax to the nearest cent, or hundredths place, before adding it to Rasheed's bill.	

Whole Numbers				·	Decimals				
One Thousands	Hundreds	Tens	Ones	Decimal Point	Tenths	Hundredths	One – Thousandths	Ten – Thousandths	Hundred – Thousandths

Underline the digit in the place value to which you are rounding.

20.5175

Look at the digit to the right of the underlined digit. If that digit is 4 or less, round down; if it is 5 or more, round up. That digit is ⬚ , so round up to 2. Leave off all digits to the right of the underlined digit.

Answer:

Example	Notes
5. Round 13.45812 to the nearest thousandth. Underline the digit in the place value to which you are rounding. 13.45812 Look at the digit to the right of the underlined digit. Round up or round down. Answer:	
7. Round 7.9561 to the nearest tenth. Answer:	

Helpful Hints

Many common statistics, for example grade point average and sports statistics are rounded to a specific place value. Many common money transactions like interest, taxes, and discounts are rounded to the nearest cent.

If the digit you are rounding up is a 9, you change that digit to a 0, and increase the digit to the left by 1. For example, 4.972 rounded to the nearest tenth is 5.0

Leave off all digits to the right of the place to which you are rounding. If you round a decimal to the nearest whole number, do not write a decimal point or any decimal places.

Concept Check

1. Which number has more decimal places, 5.325 rounded to the nearest hundredth, or 6.357895 rounded to the nearest tenth?

Practice

Round to the nearest hundredth.

2. 62.514

3. 99.999

Round to the nearest tenth.

4. 34.96

5. 0.285714

Operations with Decimals
Topic 5.4 Adding and Subtracting Decimals

Vocabulary
estimation   •   decimal places   •   decimal points   •   perimeter of a triangle

1. When adding or subtracting decimals, write the numbers vertically, making sure the _____ line up.

Step-by-Step Video Notes
Watch the Step-by-Step Video lesson and complete the examples below.

Example	Notes
1. Jennifer spent $2.97 for a meal and $1.19 for dessert, including tax. How much did she spend in total? Write the numbers vertically. Make sure the decimal points line up. 2. 9 7 +1. 1 9 □.□□ Answer:	
2. Add. Check by estimating. $3.27 + 15.2$ Add zeros as needed so all decimals have the same number of decimal places. 3 . 2 7 +1 5 . 2 □ □□.□□ Estimate. Add the whole numbers and estimate the decimal sum. The sum is about 18.5. Answer:	

Example	Notes
5. Subtract. Check by estimating.	

$21.5 - 16.43$

$$\begin{array}{r} 2\ 1\ .\ 5\ \square \\ -1\ 6\ .\ 4\ 3 \\ \hline \square\square.\square\square \end{array}$$

Answer:

6. Find the perimeter of the triangle.

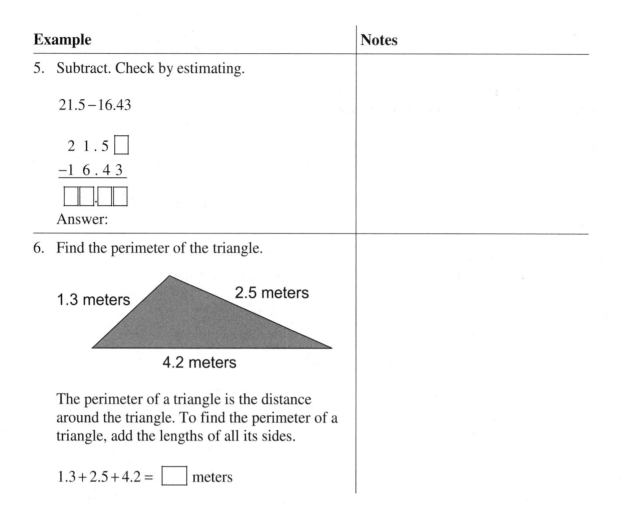

The perimeter of a triangle is the distance around the triangle. To find the perimeter of a triangle, add the lengths of all its sides.

$1.3 + 2.5 + 4.2 = \boxed{}$ meters

Helpful Hints

To add and subtract decimals, write the numbers vertically. Line up the decimal points, including the one in the answer. Add zeros to the ends of decimals as needed so all decimals have the same number of decimal places.

Concept Check

1. To subtract $24.3 - 7.52$ vertically, to which number will you insert a zero at the end?

Practice

Add.

2. $5.27 + 12.3$

3. $6 + 11.9 + 4.123$

Subtract.

4. $14.36 - 8$

5. $17.9 - 8.46$

Name: _____   Date: _____

Instructor: _____   Section: _____

Operations with Decimals
Topic 5.5 Multiplying Decimals

Vocabulary
power of 10   •   factor   •   decimal   •   whole number

1. To multiply a decimal by a _____, move the decimal point to the right the same number of places as the number of zeros in the power of 10.

Step-by-Step Video Notes
Watch the Step-by-Step Video lesson and complete the examples below.

Example	Notes

1. Multiply $3.45(6.1)$.
 	
 Count the total number of decimal places in all factors. Place the decimal point in the answer so that it has this total number of decimal places.
 	
     3 . 4 5
   ×   6. 1
 ————————
     3   4   5
 □ □ □ □ 0
 ————————
 □ □ □ □ 5
 	
 How many decimal places in all the factors? □
 	
 Answer:	

3. Multiply 3.2×0.008. 3 . 2 × 0. 0 0 8 ———————— □ . □ □ □ □ The number of decimal places needed is greater than the number of digits in the answer. Insert zeros before the answer. Answer:	

Example	Notes
5. Multiply 5.793×100. To multiply a decimal by a power of 10, move the decimal point to the right the same number of places as the number of zeros in the power of 10. Move the decimal point 2 places to the right. $5.793 \times 100 = 5\ 7\ 9\ 3$ Answer:	
8. Ian's new car travels 25.6 miles for each gallon of gas. How many miles can he travel on a full tank, which contains 12 gallons of gas? The answer will have ☐ decimal place(s). $\begin{array}{r} 2\ 5\ .\ 6 \\ \times\ 1\ \ 2 \\ \hline \end{array}$ Answer:	

Helpful Hints

After multiplying, insert zeros before the answer if the number of decimal places needed is greater than the number of digits in the answer.

When multiplying by powers of 10, if the number of zeros is more than the number of original decimal places, add zero(s) to the end of the decimal in order to move the decimal point the correct number of places.

Concept Check

1. Will you need to add zeros at the end of your answer when multiplying 24.03×100 ?

Practice

Multiply.

2. 3.2×5.08

3. 7.4×0.006

Multiply by the power of 10.

4. 8.136×100

5. 44.12×1000

Operations with Decimals
Topic 5.6 Dividing Decimals

Vocabulary
divisor   •   product   •   dividend   •   quotient

1. The _____ is the answer to the division problem.

Step-by-Step Video Notes
Watch the Step-by-Step Video lesson and complete the examples below.

Example	Notes
1. $6.3 \div 3$ Copy the decimal point in the dividend straight up into the quotient. $3\overline{)6.3}$ Divide. $3\overline{)\,6.3}$ Answer:	
3. $6.7 \div 10$ How many zeros are in 10? ☐ Move the decimal point this same number of spaces to the left; you will need to insert zeros before the whole number in order to move the decimal point the correct number of places. Answer:	

Example	Notes
5. Divide 6.1 by 3. Zeros must be added to the dividend. $\boxed{}$ $3\overline{)6.100}$ Divide until repeating pattern. Answer:	
6. $.0612 \div 0.12$ Move the decimal point to the right until the divisor is a whole number. $0.12 \rightarrow \boxed{}$ Move the decimal place the same number of places in the divisor. $0.612 \rightarrow \boxed{}$ Divide. Answer:	

Helpful Hints

To divide a decimal by a whole number, copy the decimal point straight up into the quotient.

When dividing by a decimal, the decimal point is moved the same number of places in the dividend as needed to achieve a whole number in the divisor.

Concept Check

1. What different steps are taken when dividing a decimal vs. dividing by a decimal?

Practice

Divide.

2. $8.5 \div 5$

3. $5.3 \div 4$

4. $9.4 \div 100$

5. $.0736 \div .23$

Name: _____   Date: _____

Instructor: _____   Section: _____

Operations with Decimals
Topic 5.7 Order of Operations and Decimals

Vocabulary
addition and multiplication • addition and subtraction • multiplication and division •
fraction bars

1. In the order or operations, after evaluating what is inside the parentheses and evaluating
   any exponents, the next step is to perform _____ from left to right.

Step-by-Step Video Notes
Watch the Step-by-Step Video lesson and complete the examples below.

Example	Notes
1. Simplify $3.6 + (0.2 + 0.3)^2$. Evaluate what is inside the parentheses. $(0.2 + 0.3) = \boxed{}$ Evaluate the exponent. $\left(\boxed{}\right)^2 = \boxed{}$ Perform the addition. $3.6 + \boxed{} = \boxed{}$ Answer:	
2. Simplify $(12.6 - 11.4)^2 \div (0.2)(3)$. Evaluate what is inside the parentheses. $\boxed{}$ Evaluate the exponent. $\boxed{}$ Perform the remaining operations from left to right. Answer:	

Example	Notes
3. Simplify $10.5 - 3.2 + 7.2 \div 3.6$. Since there are no parentheses or exponents begin by performing the division. $10.5 - 3.2 + \boxed{}$ Perform the remaining operations from left to right. Answer:	
5. Simplify $\dfrac{10.1 + 2.1(3)}{5.05 + 3.15}$. Answer:	

Helpful Hints
The order of operations for decimals is the same as for whole numbers.

Treat a fraction bar like there are parentheses around the numerator and the denominator.

Concept Check
1. List the order of operations.

Practice
Simplify.
2. $4.1 - (0.5 + 0.4)^2$

4. $3 + 2(0.5)^2$

3. $1.3 + 0.5 \div 0.9 - 0.4$

5. $\dfrac{0.4 + 4(0.5)}{0.2 + 0.4}$

Operations with Decimals
Topic 5.8 Converting Fractions to Decimals

Vocabulary
improper fraction　•　comparing decimals　•　repeating decimal　•　equivalent decimal

1. A decimal that repeats in a pattern without end is a(n) _____.

Step-by-Step Video Notes
Watch the Step-by-Step Video lesson and complete the examples below.

Example	Notes
1. Write $\dfrac{1}{10}$ as an equivalent decimal. What power of ten is the place value? tenths How many tenths are there? ☐ Put this answer in the tenths place value after the decimal point. Answer:	
4. Write $\dfrac{3}{5}$ as a decimal. If the denominator is not a power of 10, then divide the numerator by the denominator. Enter the denominator in the division problem. $\boxed{}\,\overline{)3}$ Perform the division. Answer:	

Example	Notes
6. Convert $\frac{2}{3}$ to a decimal. Set up the long division. $\square\overline{)\square}$ Perform the long division. $\boxed{}$ When a decimal repeats in a pattern, place a bar over the repeating number(s). Answer:	
9. Convert $\frac{11}{4}$ to a decimal. Answer:	

Helpful Hints

When converting a fraction to a decimal, divide the numerator by the denominator.

When dividing the numerator by the denominator results in a repeating pattern, placing a bar over the repeating number(s) indicates the repeating decimal.

Concept Check

1. How can a fraction with a denominator that is not power of 10 be converted to a decimal?

Practice

Convert the fraction to a decimal.

2. $\frac{42}{100}$

4. $\frac{1}{6}$

3. $\frac{5}{8}$

5. $\frac{9}{5}$

Operations with Decimals

Topic 5.9 Converting Decimals to Fractions

Vocabulary

improper fraction • denominator • numerator • simplest form

1. Write the decimal part of the decimal number as the _____ of the fraction.

Step-by-Step Video Notes

Watch the Step-by-Step Video lesson and complete the examples below.

Example	Notes
1. Write 5.277 as a fraction in simplest form. What is the place value of the last 7 in the decimal? _____ Enter this power of 10 as the denominator in the fraction and write the decimal part of the decimal number as the numerator to the fraction. $\dfrac{277}{\boxed{}}$ Write the fraction or mixed number in simplest form. Answer:	
2. Write 0.96 as a fraction in simplest form. $\dfrac{96}{\boxed{}}$ Simplify by dividing numerator and denominator by common factors. Answer:	

Example	Notes
5. Write 0.02001 as a fraction in simplest form.	
Answer:	
6. Write 4.0358 as a fraction in simplest form.	
Answer:	

Helpful Hints

The number of decimal places is equal to the number of zeros in the power of 10 in the denominator. We can also use this to find the denominator.

Always simply the fraction if possible.

Concept Check

1. What are the numerator and denominator when writing a decimal as a fraction?

Practice

Write each decimal as a fraction in simplest form.

2. 0.303   4. 6.19

3. 1.75   5. 0.492

Name: _____   Date: _____
Instructor: _____   Section: _____

Ratios, Rates, and Percents
Topic 6.1 Ratios

Vocabulary
numerator   •   denominator   •   ratio   •   fraction

1. A(n) _____ is a comparison of two like quantities, measured in the same units.

Step-by-Step Video Notes
Watch the Step-by-Step Video lesson and complete the examples below.

Example	Notes
1. A gallon of lemonade is made by mixing 2 cups of lemon juice with 14 cups of water. Write the ratio of lemon juice to water in a gallon of lemonade. Use all three forms. The ratio can be written in words. 2 to 14 The ratio can be written using a colon. 2:☐ The ratio can be written as a fraction. $\dfrac{2}{14}$ Answer:	
2. Write 5 hours to 7 hours as a ratio. Write ratio in word form. _____ Write the ratio using a colon. ☐ Write the ratio as a fraction. $\dfrac{\square}{\square}$ Answer:	

Example	Notes
4. Write 4 ounces:32 ounces as a fraction in simplest form. $$\frac{4}{\Box} = \frac{\Box}{\Box}$$ Answer:	
5. Write \$1.23 to \$3.75 as a fraction in simplest form. Answer:	

Helpful Hints

Ratios can be written in three ways: in words, using a colon, and as a fraction.

We do not need to include the units in ratios since the units are the same and can be divided out as a common factor.

Concept Check

1. How is a ratio written as a fraction?

Practice

Write the ratio in all three forms.

2. An index card measures 3 inches in width and 5 inches in length. Write the ratio of width to length.

Write ratio as a fraction in simplest form.

4. 8 inches to 12 inches

3. A classroom has 7 boys and 9 girls. Write the ratio of boys to girls in the class.

5. \$4.25 to \$6.50

Ratios, Rates, and Percents
Topic 6.2 Rates

Vocabulary

rate   •   fraction   •   ratio   •   unit rate

1. A(n) _____ is a ratio that compares two quantities with different units.

2. A(n) _____ gives the rate for one unit of the item.

Step-by-Step Video Notes
Watch the Step-by-Step Video lesson and complete the examples below.

Example	Notes
1. Write the rate of 240 miles per 10 gallons as a fraction is simplest form. The first quantity is written with units as the numerator. Write the second quantity with units as the denominator. $$\frac{240 \text{ miles}}{\boxed{} \underline{}}$$ Simplify. $\dfrac{240 \text{ miles}}{\boxed{} \underline{}} = \dfrac{\boxed{} \text{ miles}}{\boxed{} \text{ gallon}}$ Answer:	
2. Write the rate 18 tomatoes for 4 pots of stew as a fraction in simplest form. Write the ratio as a fraction include units. $$\frac{18 \underline{}}{\boxed{} \text{ pots of stew}}$$ Simplify. Answer:	

Example	Notes
4. Write the ratio $250 for 20 hours as a fraction in simplest form. Remember that units are included in a rate. Answer:	
5. A 6-pack of juice bottles costs $3.00. Find the unit rate of cost per 1 bottle. Write the rate as a fraction with units. _____ Perform the division and attach the units. Answer:	

Helpful Hints

When writing a rate as a fraction, remember that the first quantity is the numerator and the second quantity is the denominator.

To find a rate, remember to add the units into the fraction.

Concept Check
1. What is the difference between a rate and a unit rate?

Practice

Write the rate as a fraction in simplest form.
2. 320 miles per 8 hours

3. 15 gallons of paint to paint 6 apartments

Find the unit rate.
4. Carl earns $600 for 40 hours of work

5. Kim spent $450 for 15 gallons of paint.

Ratios, Rates, and Percents
Topic 6.3 Proportions

Vocabulary

ratio   •   equation   •   proportion   •   fraction

1. A(n) _____ is a statement that two quantities are the same, or equal.

2. A(n) _____ is a statement that two rates or ratios are equal.

Step-by-Step Video Notes
Watch the Step-by-Step Video lesson and complete the examples below.

Example	Notes
1. Write a proportion for 3 is to 6 as 12 is to 24.	
Write the first ratio as a fraction. $\dfrac{3}{\Box}$	
Write the second ratio as a fraction. $\dfrac{\Box}{\Box}$	
Write an equation representing that the two fractions are equal to each other.	
Answer:	
3. Determine if $\dfrac{6}{21} \overset{?}{=} \dfrac{10}{35}$ is a proportion.	
Find the cross product. $6 \cdot 35 = \Box$	
Find the cross product. $21 \cdot 10 = \Box$	
Are these cross products equal? _____	
If equal, then the statement is a proportion. If not, then the statement is not a proportion.	
Answer:	

Example	Notes
5. Find the missing number in the proportion. $$\frac{3}{9} = \frac{4}{?}$$ Replace the "?" with an n. $\frac{3}{9} = \frac{4}{n}$ Find the cross products. $3n = \boxed{}$ Solve for n by dividing both sides by 3. Answer:	
6. Find the missing number in the proportion. $$\frac{5}{6} = \frac{n}{12}$$ Answer:	

Helpful Hints

When setting up a proportion, units must "match up" or be in the same place in the fractions.

In a proportion, the cross products must be equal.

Concept Check

1. How do you determine if the statement of a fraction equal to a fraction is a proportion?

Practice

2. Write a proportion for 8 is to 12 as 2 is to 3.

3. Is this statement a proportion?
$$\frac{5}{6.2} = \frac{6}{7.3}$$

Find the missing number in the proportion.

4. $\dfrac{6}{18} = \dfrac{5}{?}$

5. $\dfrac{7}{8} = \dfrac{n}{40}$

Ratios, Rates, and Percents
Topic 6.4 Percent Notation

Vocabulary

rate   •   percent   •   denominator   •   fraction

1. _____ means per 100.

Step-by-Step Video Notes
Watch the Step-by-Step Video lesson and complete the examples below.

Example	Notes
2. Write 14% as a fraction in simplest form. Write the percent number as numerator in a fraction with denominator of 100. $\dfrac{\square}{100}$ Simplify the fraction. $\dfrac{\square}{100} = \dfrac{\square}{\square}$ Answer:	
4. Write $\dfrac{7}{100}$ as a percent. Since the denominator of the fraction is 100, write the numerator as the percent. $\square\%$ Answer:	

Example	Notes
9. Is 50% less than, equal to, or greater than one whole unit. 100% represents one whole unit. Answer:	
13. A company produced 2500 personal computers, and 3% of them were defective. How many of the computers were defective? To find 3% of 2500, turn 3% into a fraction. $3\% = \dfrac{\Box}{\Box}$ Multiply this fraction by 2500. Answer:	

Helpful Hints

To convert p% to a fraction, write $\dfrac{p}{100}$.

Remember that 100% represents one whole unit.

Concept Check
1. How is the percent of a number found?

Practice
2. Write 128% as a fraction in simplest form.

4. Is 225% less than, equal to, or greater than one whole unit?

3. Write $\dfrac{29}{100}$ as a percent.

5. A company has 600 employees, and 52% of them are male. How many employees are male?

Ratios, Rates, and Percents
Topic 6.5 Percent and Decimal Conversions

Vocabulary

decimal   •   percent   •   ratio   •   fraction

1. _____ means per 100.

Step-by-Step Video Notes
Watch the Step-by-Step Video lesson and complete the examples below.

Example	Notes
1. Write 51% as a decimal. Place the decimal at the end of the number and then move it two places to the left. [.] The percent sign is left off. Answer:	
5. Write 4.5% as a decimal. Answer:	
6. Write 0.71 as a percent. Move the decimal two places to the right, adding zeros if needed. 0.71 → [ ] Write the percent sign. Answer:	

Example	Notes
9. Write 0.75 as a percent.	
Answer:	

Helpful Hints
When converting a percent to a decimal, it may be necessary to add zeros.

When converting a decimal to a percent, it may be necessary to add zeros.

Concept Check
1. What are the similarities and differences in converting a percent to a decimal and converting a decimal to a percent?

Practice
Write the percent as a decimal.

2. 32.5%

Write the decimal as a percent.

4. 1.8

3. 6.1%

5. 0.025

Ratios, Rates, and Percents
Topic 6.6 Percent and Fraction Conversions

Vocabulary

decimal   •   percent   •   ratio   •   fraction

1. If the denominator of a fraction is 100, then the numerator is the _____.

Step-by-Step Video Notes
Watch the Step-by-Step Video lesson and complete the examples below.

Example	Notes
1. Write 8% as a fraction in simplest form. Write the percent as the numerator with the denominator as 100. $$\dfrac{\Box}{100}$$ Simplify. Answer:	
4. The sales tax is 4.5% of the price. Write the percent as a fraction in simplest form. Write the percent as an equivalent fraction. $$4.5 = \dfrac{\Box}{10}$$ Multiply this fraction by $\dfrac{1}{100}$. $$\dfrac{\Box}{10} \cdot \dfrac{1}{100} = \dfrac{\Box}{\Box}$$ Answer:	

Example	Notes
6. Write $\dfrac{5}{8}$ as a percent. Divide 5 by 8. Convert to a percent by moving the decimal place two places to the right and then write the percent sign. Answer:	
7. Write $\dfrac{4}{7}$ as a percent. Round to the nearest tenth of a percent. Answer:	

Helpful Hints

You can convert p% to a fraction by dividing p by 100 or by multiplying p by $\dfrac{1}{100}$.

Remember that when a fraction has a denominator of 100, the numerator is the percent.

Concept Check
1. What are the similarities and differences in converting a percent to a fraction and converting a fraction to a percent?

Practice

Write the % as a fraction in simplest form.

2. 12%

3. 6.1%

Write the fraction as a decimal. Round to the nearest tenth of a percent as needed.

4. $\dfrac{3}{5}$

5. $\dfrac{5}{6}$

Ratios, Rates, and Percents
Topic 6.7 The Percent Equation

Vocabulary
decimal   •   percent equation   •   base   •   fraction

1. The _____ states "the amount is equal to the percent times the base."

Step-by-Step Video Notes
Watch the Step-by-Step Video lesson and complete the examples below.

Example	Notes
1. What amount is 30% of 140? Substitute the values into the percent equation, $amount = \% \times base$. Enter 30% as a decimal. $a = \boxed{} \times 140$ Multiply to find a. Answer:	
3. 34 is 50% of what number? Substitute the values into the percent equation, $amount = \% \times base$. Enter 50% as a decimal. $34 = \boxed{} \times b$ Solve the equation for b by dividing each side of the equation by the decimal in front of b. Answer:	

Example	Notes

5. 45 is what percent of 180?
 	
   Substitute the values into the percent equation, amount = % × base , letting p stand for the unknown percent.
 	
   $\boxed{} = \text{p} \times \boxed{}$
 	
   Solve for p.
 	
   Answer:	

6. What percent of 30 is 10? Round to the nearest tenth of a percent. Answer:	

Helpful Hints
Usually the base appears after the word "of."

In math "is" translates to an equal sign, and "of" translates into multiplication.

Concept Check
1. State the percent equation and how it can be used to find each of one of the three parts, provided the other two parts are given.

Practice
Use the percent equation to find the following. Round to the nearest tenth of a percent as needed.

2. What amount is 40% of 150?

4. 32 is what percent of 160?

3. 24 is 20% of what number?

5. 25 is what percent of 86?

Ratios, Rates, and Percents
Topic 6.8 The Percent Proportion

Vocabulary
amount   •   percent proportion   •   ratio   •   fraction

1. The _____ states "the ratio of the amount to the base is equal to the ratio of the

   percent $\dfrac{p}{100}$."

Step-by-Step Video Notes
Watch the Step-by-Step Video lesson and complete the examples below.

Example	Notes
1. What amount is 20% of 85? Substitute the values into the percent proportion, $\dfrac{amount}{base} = \dfrac{p}{100}$. $\dfrac{a}{\boxed{}} = \dfrac{20}{100}$ Cross multiply. $100(a) = \boxed{}$ Solve for a by dividing each side of the equation by 100. Answer:	
2. 21 is 10.5% of what number? Substitute the values into the percent proportion, $\dfrac{amount}{base} = \dfrac{p}{100}$. $\dfrac{\boxed{}}{b} = \dfrac{\boxed{}}{100}$ Cross multiply. $\left(\boxed{}\right)b = \left(\boxed{}\right)(100)$ Solve for b. Answer:	

Example	Notes
3. 83 is what percent of 332? Substitute the values into the percent proportion, $\dfrac{\text{amount}}{\text{base}} = \dfrac{p}{100}$. Solve for p. Answer:	
4. 4 is what percent of 48? Round to the nearest tenth of a percent. Answer:	

Helpful Hints

The percent proportion is an alternative to the percent equation.

Usually the base appears after the word "of."

Concept Check

1. State the percent proportion and how it can be used to find each of one of the three parts, provided the other two parts are given.

Practice

Use the percent proportion to find the following.

2. What amount is 30% of 160?

3. 33 is 11% of what number?

Round to the nearest tenth as necessary.

4. 96 is what % of 384?

5. 6 is what percent of 96?

Ratios, Rates, and Percents
Topic 6.9 Percent Applications

Vocabulary
amount   •   percent   •   base   •   fraction

1. The percent equation states "the amount is equal to the percent times the _____."

2. The percent proportion states "the ratio of the amount to the _____ is equal to the ratio of the percent $\dfrac{p}{100}$."

Step-by-Step Video Notes
Watch the Step-by-Step Video lesson and complete the examples below.

Example	Notes
1. Eli purchased a wrist watch for \$60. If the sales tax rate is 7%, how much sales tax did Eli pay? What was the total cost of the watch? Substitute the values into the application percent equation: $\left(\begin{array}{c}\text{amount of}\\\text{sales tax}\end{array}\right)=\left(\begin{array}{c}\text{tax rate}\\\text{as a decimal}\end{array}\right)\times\left(\begin{array}{c}\text{cost of}\\\text{item}\end{array}\right)$ $a=(.07)\times\boxed{}$ Multiply to solve for a. $a=\boxed{}$ Substitute values and perform addition to find total cost. $\left(\begin{array}{c}\text{total cost}\\\text{of item}\end{array}\right)=\left(\begin{array}{c}\text{amount of}\\\text{sales tax}\end{array}\right)+\left(\begin{array}{c}\text{cost of}\\\text{item}\end{array}\right)$ Answers:	

Example	Notes
4. Calculate the amount of interest and the total amount paid for a loan of $2000 for 3 years at 4% annual rate interest. First we calculate the amount of interest: $\text{Amount of Interest} = \text{Principal} \times \text{Rate} \times \text{Time}$ Next we calculate the total amount paid: $\text{Total Amount} = \text{Principal} + \text{Amount of Interest}$ Answer:	

Helpful Hints

If you know the percent and the base when solving a percent application problem, you should use the percent equation.

Remember to use the percent as a decimal in the percent equation.

Concept Check

1. How does finding the amount of sales tax of an item make use of the percent equation?

Practice

Samantha wants to buy a $40 sweater. Find the following.

2. What is the amount of sales tax if the sales tax rate is 5%?

4. What is the amount of discount if there is a 15% discount sale?

3. What is the total cost of the sweater if the sales tax rate is 5%

5. What is the sale price if Samantha buys the sweater during the 15% discount sale?

U.S. and Metric Measurement
Topic 7.1 U.S. Length

Vocabulary
conversion   •   unit fraction   •   length   •   mixed unit

1. A _____ has a value of 1. Its numerator and denominator express the same value in different ways.

Step-by-Step Video Notes
Watch the Step-by-Step Video lesson and complete the examples below.

Example	Notes
1. Convert 24 yards to feet using unit fractions.	

Table 2 U.S. Length Conversions
12 in. = 1ft
36 in. = 1 yd
3 ft = 1 yd
5280 ft = 1 mi
1760 yd = 1 mi

There are ☐ ft in 1 yd. Using

$\dfrac{\text{unit of measurement converting to}}{\text{original unit of measurement}}$, the

numerator is in ft, and the denominator is in yd.

24 yards $\cdot \dfrac{\boxed{} \text{ feet}}{1 \text{ yard}} = \boxed{}$ feet

2. Convert 15,840 ft to mi using unit fractions.

There are $\boxed{}$ feet in 1 mile.

15,840 ft $\cdot \dfrac{1 \text{ mi}}{\boxed{} \text{ ft}} = \boxed{}$ mi

Example	Notes
4. Convert the following using unit fractions. 90 in. to yd 90 in. $\cdot \dfrac{\boxed{}}{\boxed{}} \rule{2cm}{0.4pt} = \boxed{}$ yd Answer:	
5. Gage is 54 inches tall. Convert this to feet and inches. 54 in. $\cdot \dfrac{\boxed{}}{\boxed{}} \rule{2cm}{0.4pt} = \boxed{}$ r $\boxed{}$ The quotient is the number of feet, and the remainder is in inches. Answer:	

Helpful Hints

Use unit fractions such as $\dfrac{12 \text{ in.}}{1 \text{ ft}}$, $\dfrac{36 \text{ in.}}{1 \text{ yd}}$, $\dfrac{3 \text{ ft}}{1 \text{ yd}}$, $\dfrac{5280 \text{ ft}}{1 \text{ mi}}$, and $\dfrac{1760 \text{ yd}}{1 \text{ mi}}$ to convert units.

The fraction $\dfrac{\text{unit of measurement converting to}}{\text{original unit of measurement}}$ can be helpful to use when converting units.

Concept Check
1. Why is a unit fraction equal to 1 if the numerator and denominator have different numbers?

Practice

Convert the following using unit fractions.
2. 468 in. to yd

3. 3.1 miles to yards

Convert the heights to feet and inches.
4. Ben is 68 inches tall.

5. Gomez is 76 inches tall.

U.S. and Metric Measurement
Topic 7.2 U.S. Weight and Capacity

Vocabulary
weight   •   mass   •   capacity   •   pound   •   gallon

1. _____ is related to the gravitational pull on an object.

2. _____ is the amount of space inside a three-dimensional figure.

Step-by-Step Video Notes
Watch the Step-by-Step Video lesson and complete the examples below.

Example	Notes
1. Convert 7.5 tons to pounds.	

U.S. Weight Conversions

16 oz = 1 lb
2000 lbs = 1 ton

There are ☐ pounds in 1 ton.

Using $\dfrac{\text{unit of measurement converting to}}{\text{original unit of measurement}}$, the

numerator is in pounds, and the denominator is in tons.

7.5 tons $\cdot \dfrac{\boxed{}\ \text{pounds}}{1\ \text{ton}} = \boxed{}$ pounds

Answer:

2. Convert 64 oz to lb.

There are ☐ oz in 1 lb.

64 oz $\cdot \dfrac{1}{\boxed{}\ \text{oz}} = \boxed{}\ \boxed{}$

Answer:

Example	Notes
4. Convert 26 quarts to gallons.	

<table>
<tr><td colspan="2">Table 4 U.S. Capacity Conversions</td></tr>
<tr><td colspan="2">8 fluid ounces = 1 cup</td></tr>
<tr><td colspan="2">2 cups = 1 pint</td></tr>
<tr><td colspan="2">16 fluid ounces = 1 pint</td></tr>
<tr><td colspan="2">2 pints = 1 quart</td></tr>
<tr><td colspan="2">4 quarts = 1 gallon</td></tr>
</table>

Give your answer in decimal or fraction form.

Answer:

5. Lashonda buys a bottle of ketchup that contains 44 fl oz. Convert this to pints and fluid ounces.

$$44 \text{ fl oz.} \cdot \frac{\Box}{\Box}\frac{\rule{2cm}{0.4pt}}{\rule{2cm}{0.4pt}} = \Box \text{ r } \Box$$

The quotient is the number of pints, and the remainder is in fluid ounces.

Answer:

Helpful Hints

Use unit fractions such as $\dfrac{16 \text{ oz}}{1 \text{ lb}}, \dfrac{2000 \text{ lb}}{1 \text{ ton}}, \dfrac{8 \text{ fl oz}}{1 \text{ cup}}, \dfrac{16 \text{ fl oz}}{1 \text{ pint}}, \dfrac{1 \text{ quart}}{2 \text{ pints}},$ and $\dfrac{4 \text{ quarts}}{1 \text{ gallon}}$ to convert units. Note that weight and capacity are often given in mixed units, such as pounds and ounces, pints and fluid ounces, etc.

Concept Check
1. Why use pounds and ounces for the weight of a newborn baby, rather than a decimal number, as is often used with weights measured in tons?

Practice
Convert the following to pounds.
2. 96 ounces

3. 3.7 tons

Convert the following to gallons.
4. 64 fl oz

5. 36 quarts

U.S. and Metric Measurement
Topic 7.3 Metric Length

Vocabulary
metric prefixes   •   meter   •   milli-   •   centi-   •   kilo-
deka-   •   hecto-   •   deci-

1. The metric prefix _____ means 1000.

Step-by-Step Video Notes
Watch the Step-by-Step Video lesson and complete the examples below.

Example	**Notes**
1. Convert 400 centimeters to meters .	

Table 1 Metric Units of Length	
Conversion	**Unit Fraction**
1000 millimeters (mm) = 1 meter (m)	$\dfrac{1 \text{ meter}}{1000 \text{ millimeters}}$ or $\dfrac{1000 \text{ millimeters}}{1 \text{ meter}}$
100 centimeters (cm) = 1 meter (m)	$\dfrac{1 \text{ meter}}{100 \text{ centimeters}}$ or $\dfrac{100 \text{ centimeters}}{1 \text{ meter}}$
10 decimeters (dm) = 1 meter (m)	$\dfrac{1 \text{ meter}}{10 \text{ decimeters}}$ or $\dfrac{10 \text{ decimeters}}{1 \text{ meter}}$
1 meter (m) is the basic unit of length	
10 meters (m) = 1 dekameter (dam)	$\dfrac{10 \text{ meter}}{1 \text{ dekameters}}$ or $\dfrac{1 \text{ dekameters}}{10 \text{ meter}}$
100 meters (m) = 1 hectometer (hm)	$\dfrac{100 \text{ meter}}{1 \text{ hectometer}}$ or $\dfrac{1 \text{ hectometer}}{100 \text{ meter}}$
1000 meters (m) = 1 kilometer (km)	$\dfrac{1000 \text{ meter}}{1 \text{ kilometer}}$ or $\dfrac{1 \text{ kilometer}}{1000 \text{ meter}}$

There are ☐ centimeters in 1 meter.

$$400 \text{ centimeters} \cdot \frac{1 \text{ meter}}{100 \text{ centimeters}} = \boxed{} \text{ meters}$$

Answer:

Example	Notes

3. Convert 96.3 km to m.
 	
   List the prefixes. <u>km hm dam m</u> dm cm mm
 	
   To convert from km to m, move the decimal point ☐ places to the _____.
 	
   Answer:	

4. Convert 150 dam to cm.
 	
   150 dam = ☐☐☐,☐☐☐ cm
 	
   Answer:	

Helpful Hints

The metric system is based on powers of 10. Converting units can be done by moving the decimal point. The mnemonic "Kangaroos hopping down mountains drinking chocolate milk" can help you remember the metric prefixes in order from largest to smallest, kilometers, hectometers, decameters, meter, decimeter, centimeter, millimeter.

List the prefixes like this km hm dam m dm cm mm is a visual way to tell which way to move the decimal point when converting metric units of length. Start with the original unit and move to the new unit. Move the decimal point accordingly, the same number of spaces and in the same direction, adding zeros as necessary.

Concept Check

1. When converting from meters to centimeters, how many places and in what direction should you move the decimal point?

Practice

Convert the following to meters.

2. 87 cm

3. 6.2 km

Convert the following to centimeters.

4. 25 m

5. 44 mm

U.S. and Metric Measurement
Topic 7.4 Metric Mass and Capacity

Vocabulary

gram   •   mass   •   kilogram   •   milligram   •   dekagram
liter   •   milliliter   •   capacity   •   deciliter   •   kiloliter

1. The basic unit of mass in the metric system is the _____.

2. A _____ is slightly more than two pounds.

Step-by-Step Video Notes
Watch the Step-by-Step Video lesson and complete the examples below.

Example	Notes
1. Convert 32 centigrams to grams. There are ☐ centigrams in 1 gram. Using $\dfrac{\text{unit of measurement converting to}}{\text{original unit of measurement}}$, the numerator is in grams, and the denominator is in centigrams. $32 \text{ cg} \cdot \dfrac{1 \text{ g}}{\boxed{} \text{ cg}} = \boxed{} \text{ g}$ Answer:	
3. Convert 216 kg to cg. List the prefixes. kg hg dag g dg cg mg To convert from kg to cg, move the decimal point ☐ places to the _____. Add zero(s) to the end of the decimal to move the decimal point the correct number of places. Answer:	

Example	Notes
4. Convert 900 mL to L. List the prefixes. kL hL daL <u>L</u> dL cL mL To convert from mL to L, move the decimal point ☐ places to the _____ . Answer:	
5. Convert 83.2 L to cL. Answer:	

Helpful Hints
Regardless of the type of measure (length, mass, or capacity), the metric prefixes always have the same meaning and relationship to the basic unit.

With mass and capacity, the prefixes kilo- and milli- are most often used.

Concept Check
1. Which metric unit is closest to a quart in U.S. measurement? Is a gallon more or less than 4 liters?

Practice

Convert the following to grams.
2. 96 mg

3. 5.8 kilograms

Convert the following to liters.
4. 500 mL

5. 4.4 kL

U.S. and Metric Measurement
Topic 7.5 Converting between U.S. and Metric Units

Vocabulary
meter   •   gallon   •   pound   •   approximately   •   exactly

1. The symbol ≈ means _____.

Step-by-Step Video Notes
Watch the Step-by-Step Video lesson and complete the examples below.

Example	Notes
1. Convert 44.02 miles to kilometers.	

Table 1 U.S. to Metric (Length)	
Conversion	**Unit Fraction**
1 inch (in.) = 2.54 centimeters (cm)	$\dfrac{2.54 \text{ cm}}{1 \text{ in.}}$
1 foot (ft) = 0.30 meter (m)	$\dfrac{0.30 \text{ m}}{1 \text{ ft}}$
1 yard (yd) = 0.91 meter (m)	$\dfrac{0.91 \text{ m}}{1 \text{ yd}}$
1 mile (mi) = 1.61 kilometers (km)	$\dfrac{1.61 \text{ km}}{1 \text{ mi}}$

Table 2 Metric to U.S. (Length)	
Conversion	**Unit Fraction**
1 meter (m) = 39.37 inches (in.)	$\dfrac{39.37 \text{ in.}}{1 \text{ m}}$
1 meter (m) = 1.09 yards (yd)	$\dfrac{1.09 \text{ yd}}{1 \text{ m}}$
1 meter (m) = 3.28 feet (ft)	$\dfrac{3.28 \text{ ft}}{1 \text{ m}}$
1 kilometer (km) = .62 mile (mi)	$\dfrac{62 \text{ mi}}{1 \text{ km}}$

There are about ⬜ kilometers in 1 mile.

Convert using the unit fraction.

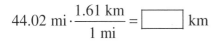

$44.02 \text{ mi} \cdot \dfrac{1.61 \text{ km}}{1 \text{ mi}} = \boxed{} \text{ km}$

Answer: _____

Example	Notes

3. Convert 5.2 kg to lb. Use the unit fraction

Table 3 U.S. to Metric (Weight/Mass)	
Conversion	**Unit Fraction**
1 ounce (oz) = 28.35 grams (g)	$\dfrac{28.35\ g}{1\ oz}$
1 pound (lb) = 0.45 kilogram (kg)	$\dfrac{0.45\ kg}{1\ lb}$

Table 4 Metric to U.S. (Weight/Mass)	
Conversion	**Unit Fraction**
1 kilogram (kg) = 2.20 pounds (lb)	$\dfrac{2.20\ lb}{1\ kg}$
1 gram (g) = 0.035 ounce (oz)	$\dfrac{0.035\ oz}{1\ g}$

Answer:

6. Convert 3 qt to L. Use the unit fraction to convert. Round to the nearest tenth.

Table 5 U.S. to Metric (Capacity)	
Conversion	**Unit Fraction**
1 quart (qt) = 0.95 liter (L)	$\dfrac{0.95\ L}{1\ qt}$
1 gallon (gal) = 3.79 liters (L)	$\dfrac{3.79\ L}{1\ gal}$

Table 6 Metric to U.S. (Capacity)	
Conversion	**Unit Fraction**
1 liter (L) = 1.06 quarts (qt)	$\dfrac{1.06\ qt}{1\ L}$
1 liter (L) = 0.26 gallon (gal)	$\dfrac{0.26\ gal}{1\ L}$

Answer:

Helpful Hints
Almost all unit fractions used to convert between U.S. and metric units are approximate. Not all tables provide every conversion fact. You may need to change to units provided in the table, then use another conversion fact you know to complete the conversion.

Concept Check
1. When measuring a length, will there be more units if you measure in yards or in meters?

Practice
Convert the following to meters.
2. 100 yards
3. 6.2 miles

Convert the following to quarts.
4. 18 L
5. 947 mL

U.S. and Metric Measurement
Topic 7.6 Time and Temperature

Vocabulary
conversion   •   unit fraction   •   denominator   •   equivalent fraction

1. A _____ has a value of 1. Its numerator and denominator express the same value in different ways.

Step-by-Step Video Notes
Watch the Step-by-Step Video lesson and complete the examples below.

Example	Notes
1. Convert 195 minutes to hours. Use the unit fraction, where the numerator is hours and the denominator is minutes: $$\frac{\text{unit of measurement converting to}}{\text{original unit of measurement}}$$ Fill in with the appropriate values. $$\frac{1 \text{ hour}}{\boxed{} \text{ minutes}}$$ Multiply 195 minutes by the unit fraction. Answer:	
2. Convert 4.4 hours to seconds. First convert hours to minutes using the unit fraction. $$4.4 \text{ hours} \times \frac{\boxed{} \text{ minutes}}{1 \text{ hour}} = \boxed{} \text{ minutes}$$ Next convert these minutes to seconds. Answer:	

Example	Notes
3. Convert 36° Fahrenheit to degrees Celsius. Use the following formula: $C = \dfrac{5 \times F - 160}{9}$ Fill in degrees Fahrenheit. $C = \dfrac{5 \times \left(\square\right) - 160}{9}$ Solve for C by following order of operations. Answer:	
4. Convert 81° Celsius to degrees Fahrenheit. Round to the nearest tenth. Use the following formula: $F = 1.8 \times C + 32$ Answer:	

Helpful Hints

The fraction $\dfrac{\text{unit of measurement converting to}}{\text{original unit of measurement}}$ is helpful when converting units of time.

When converting units of time, sometimes more than one unit fraction will be needed.

Concept Check
1. What unit fraction would you multiply by to convert days to minutes?

Practice

Convert the following measurements of time using unit fractions.
2. 145 minutes to hours

Convert the following measurements of temperature using the appropriate formula.
4. 100° C to degrees Fahrenheit

3. 1.5 hours to seconds

5. 72° F to degrees Celsius

Name: _____   Date: _____
Instructor: _____   Section: _____

Introduction to Geometry
Topic 8.1 Lines and Angles

Vocabulary
point • line • line segment • ray • angle • measurement • right angle
acute angle • straight angle • obtuse angle • complementary • supplementary

1. A(n) _____ is a portion of a line with two endpoints.

2. A(n) _____ is a portion of a line with one endpoint.

3. A(n) _____ is an angle that measures 90°.

4. Two angles are _____ if the sum of their measures is 180°.

Step-by-Step Video Notes
Watch the Step-by-Step Video lesson and complete the examples below.

Example	Notes
1. Identify the following figure as a line, line segment, or a ray and give the name.  A●————————●————————▶ B Use the endpoint(s) and the appropriate symbol to name the figure. Answer:	
4b. Identify the angle shown below as a straight angle, right angle or neither. 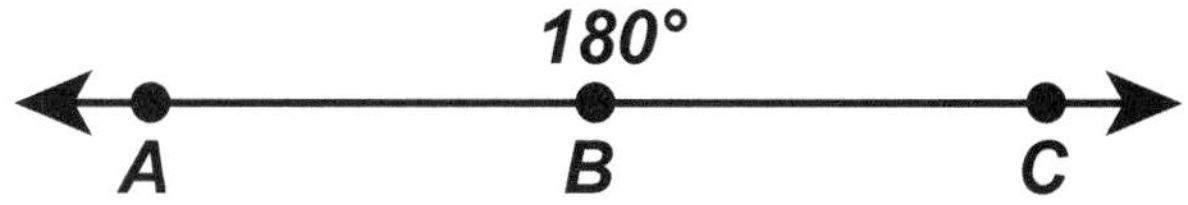 **180°** ◀————●————————●————————●————▶ A B C Answer:	

Example	Notes
5a. Identify the given angle as acute, obtuse, or neither. ![angle ABC diagram] Review the definitions of acute angle and obtuse angle. Answer:	
7a. Find the complement of a 65° angle. Two angles are complementary if the sum of their angles is ▢°. Subtract 65° from the sum. Answer:	

Helpful Hints

If two complementary angles are adjacent, they will form a right angle.

If two supplementary angles are adjacent, they will form a straight angle.

Concept Check

1. Describe how 90° and 180° are used to define the following angles: acute, right, obtuse, straight, complementary and supplementary.

Practice

Identify the type of angle described as acute, right, or obtuse.

2. An angle measuring 65°

3. An angle measuring 90°

Find the following angles.

4. The supplement of 120°

5. The complement of 35°

Introduction to Geometry
Topic 8.2 Figures

Vocabulary

triangle	•	right triangle	•	acute triangle	•	obtuse triangle	
polygon	•	quadrilateral	•	parallel lines	•	rectangle	
square	•	trapezoid	•	parallelogram	•	parallel lines	

1. A(n) _____ is a four-sided geometric figure.

2. A(n) _____ is a quadrilateral with opposite sides that are equal in length and four angles that are right angles.

3. A(n) _____ is a quadrilateral with only one pair of opposite sides that are parallel.

Step-by-Step Video Notes
Watch the Step-by-Step Video lesson and complete the examples below.

Example	Notes
1. If two angles of a triangle measure 55° and 70°, find the measure of the third angle. Then identify the triangle as acute, right, or obtuse. Find the sum of the two given angles. ☐° Subtract this sum from 180. $180 - \boxed{} = \boxed{}$ Is there an obtuse angle? _____ Is there a right angle? _____ Are all the angles acute? _____ Answer:	

Example	Notes
2. Identify the figure below. ![triangle] Answer:	
3. Identify the figure below. ![rectangle with right angle mark] Answer:	

Helpful Hints
The sum of the angles of any triangle is 180°.

Every square is also a rectangle, but every rectangle is not a square.

Concept Check
1. What are the similarities and differences among these quadrilaterals: rectangle, square, trapezoid, parallelogram, diamond and kite.

Practice
2. Find the third angle of a triangle having angles of 100° and 50°.

4. Identify a triangle with angles of 95°, 45°, and 40° as acute, right or obtuse.

3. Identify the figure below.

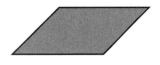

5. Identify the figure below.

Introduction to Geometry
Topic 8.3 Perimeter – Definitions and Units

Vocabulary
polygon   •   distance   •   perimeter   •   parallelogram

1. The _____ of a figure can be found by adding the lengths of all its sides.

Step-by-Step Video Notes
Watch the Step-by-Step Video lesson and complete the examples below.

Example	Notes
1. Find the perimeter of the figure. 6 inches 8 inches 4 inches Add the lengths of the three sides of the figure. 6 inches + ☐ inches + ☐ inches = ☐ inches Answer:	
2. Find the perimeter of the figure. 3 cm 3 cm 3 cm 3 cm 3 cm Add the lengths of the five sides of the figure. Answer:	

Example	Notes
3. Find the perimeter of the figure.	

0.5 meters

4 meters

4.5 meters

2.5 meters

0.5 meters

3 meters

Answer:

Helpful Hints

Perimeter is always measured in units of length.

Concept Check

1. What is the similarity in finding the perimeter of a rectangle and a triangle?

Practice

2. Which of these units, if any, could represent perimeter? Square feet, miles, or cubic meters

4. Find the perimeter of the figure.

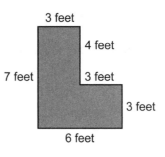

3 feet

4 feet

7 feet

3 feet

3 feet

6 feet

3. Find the perimeter of the figure.

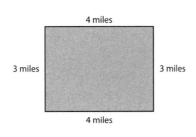

4 miles

3 miles   3 miles

4 miles

5. Find the perimeter of the figure.

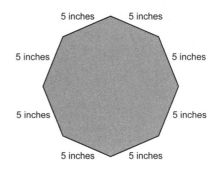

5 inches   5 inches

5 inches   5 inches

5 inches   5 inches

5 inches   5 inches

Introduction to Geometry
Topic 8.4 Finding Perimeter

Vocabulary

angle   •   square   •   perimeter   •   rectangle

1. The _____ of a figure can be found by adding the lengths of all its sides.

Step-by-Step Video Notes
Watch the Step-by-Step Video lesson and complete the examples below.

Example	**Notes**
1. Find the perimeter of the triangle below. 3.5 cm 2.5 cm 4 cm Use the perimeter formula: $P = a + b + c$, where a, b, and c represent the sides of the triangle. $P = 3.5 \text{ cm} + \boxed{} \text{ cm} + \boxed{} \text{ cm}$ Find the sum. Answer:	
2. Find the perimeter of the rectangle below. 30 feet 40 feet Use the perimeter formula: $P = 2l + 2w$, where l represents length and w represents width. Answer:	

Example	Notes
3. Find the perimeter of the figure. 14 m 14 m Use the formula: $P = 4s$, where s represents the side length of the square. Answer:	
4. Find the perimeter of a triangular garden with side lengths of 2 feet, 5 feet, and 10 feet. Answer:	

Helpful Hints

Perimeter is always measured in units of length.

The perimeter can be found by adding the lengths of the sides of a figure, but formulas for a triangle, a square and a rectangle can be used.

Concept Check

1. What are the perimeter formulas for a triangle, a rectangle, and a square?

Practice

Use the appropriate perimeter formula to find the perimeter of the figures shown or described below.

2.

6 cm   6 cm

10 cm

4.

4 inches

12 inches

3. A rectangular desk with length 4 feet and width 2.5 feet.

5. A square bandana with side lengths of 8 inches.

Introduction to Geometry
Topic 8.5 Area – Definitions and Units

Vocabulary

square unit   •   area   •   perimeter   •   square inches

1. The _____ of a figure is the measure of the surface inside the figure, which is measured in square units.

Step-by-Step Video Notes
Watch the Step-by-Step Video lesson and complete the examples below.

Example	Notes
1. How many square units are needed to cover the figure below completely? 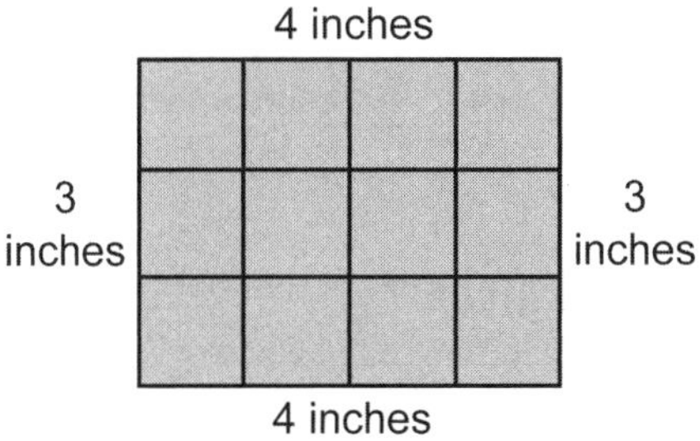 4 inches 3 inches · · · 3 inches 4 inches The small squares measuring 1 in.×1 in. are square units. Count the number of square units in this figure. Answer:	
3. Draw two different rectangles each of which has an area of 6 square units. 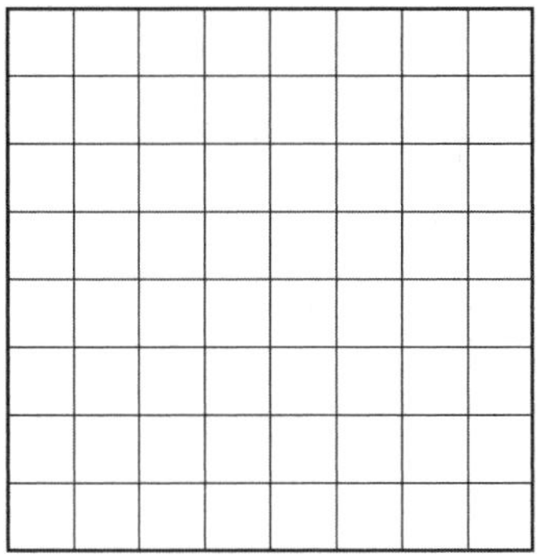	

Example	Notes

5. Find the area of a right triangle with a base and height of 4 in.

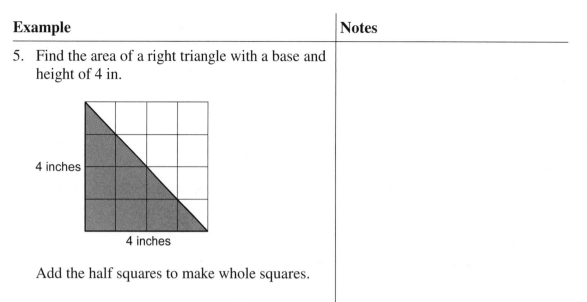

4 inches

4 inches

Add the half squares to make whole squares.

Answer:

Helpful Hints

Area is always measured in square units of length.

Two figures can have different shapes, but have the same area.

Concept Check

1. How is a grid of square units used to determine the area of a figure?

Practice

6. Find the area of a right triangle with a base and height of 3 in.

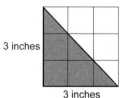

3 inches

3 inches

3. How many square units are needed to cover the figure below completely?

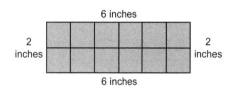

6 inches

2 inches   2 inches

6 inches

4. Draw a rectangle which has an area of 12 square units.

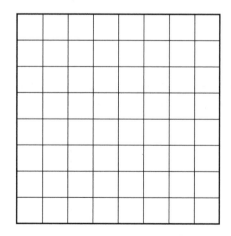

Introduction to Geometry
Topic 8.6 Finding Area

Vocabulary
square unit   •   area   •   perimeter   •   square inches

1. The _____ of a figure is the measure of the surface inside the figure, which is measured in square units.

Step-by-Step Video Notes
Watch the Step-by-Step Video lesson and complete the examples below.

Example	Notes
1. Use the formula for area of a rectangle to find the area of the rectangular photo shown below. 9 inches 6 inches $A = lw$, where l is length and w is width $A = 9 \text{ in.}\left(\boxed{} \text{ in.}\right) = \boxed{}$ square inches Answer:	
2. Use the formula for area of a square to find the area of a square deck with sides 3 meters. 3 m 3 m $A = s^2$, where s is the side $A = \left(\boxed{} \text{ meters}\right)^2 = \boxed{}$ square meters Answer:	

Example	Notes
3. Use the formula for area of a triangle to find the area of a right triangle with a base of 4 cm and a height of 3 cm. 3 cm ◺ 4 cm Answer:	
4. Find the area of the infield of a major league baseball diamond which is a square whose sides are 90 feet long. 90 feet 90 feet 90 feet 90 feet Answer:	

Helpful Hints

The basic shapes have formulas which can be used to find their area.

Remember to include the square units of length when finding area of a shape.

Concept Check

1. What are the formulas for finding the area of a rectangle, a square, and a right triangle?

Practice

Find the area of the following.

2. A rectangle whose length is 12 inches and width is 3 inches

3. A square whose sides are 7 meters long

4. A right triangle with a base of 12 cm and a height of 5 cm

5. A diamond logo on a shirt which is a square whose sides are 8 cm long.

Introduction to Geometry
Topic 8.7 Understanding Circles

Vocabulary
circle • diameter • area • radius • circumference • pi

1. A _____ is a figure in which all points on the circle are the same distance from a fixed point called the center.

2. The _____ of a circle is the distance from the center to a point on the circle.

3. The _____ of a circle is the distance around the circle.

Step-by-Step Video Notes
Watch the Step-by-Step Video lesson and complete the examples below.

Example	Notes
1. Use the appropriate formula to find the radius of a pizza with a diameter of 10 inches. Use the formula $r = \dfrac{1}{2}d$, where d is the diameter. $r = \dfrac{1}{2}\left(\boxed{}\text{ inches}\right) = \boxed{}\text{ inches}$ Answer:	
2. A neighbor purchased a 14-foot circular trampoline for his children. Does this situation involve the diameter or the radius? 14 foot Review the definitions for diameter and radius. Answer:	

Example	Notes
3. In softball, the circle around the pitcher's mound is drawn 6 feet from where the pitcher stands. Does this situation involve the diameter or the radius? 6 feet Answer:	

Helpful Hints
The radius is always smaller than the diameter; the radius is half of the diameter.

The radius extends from the center of a circle, while the diameter passes through the center.

Concept Check
1. Define these parts of a circle: radius, diameter and circumference.

Practice
Does the situation described involve the diameter of the radius?

2. A toy helicopter has moveable blades that are 5 inches long.

Find the radius in the following situations.

4. A circular garden with a diameter of 4 feet.

3. A child's circular board game folds along the center with a measure of 15 inches.

5. A circular drum with a diameter of 12 inches.

Introduction to Geometry
Topic 8.8 Finding Circumference

Vocabulary
diameter   •   pi   •   radius   •   circumference   •   area

1. $\pi = \dfrac{C}{d}$, where C is the circumference and d is the _____.

2. The _____ of a circle is the distance around the circle.

Step-by-Step Video Notes
Watch the Step-by-Step Video lesson and complete the examples below.

Example	Notes
1. Find the circumference of a circle with a radius of 8 inches. Find the exact answer in terms of π. Then find the approximation using 3.14 for π. Use the formula $C = 2\pi r$ $C = 2\pi \left(\boxed{}\ \text{in.}\right) = \boxed{}\ \pi\ \text{in.}$ Use 3.14 for π. $C = \boxed{}(3.14)\ \text{in.} = \boxed{}\ \text{in.}$ Answer:	
2. Find the circumference of a circle with a diameter of 5 cm. Find the exact answer in terms of π. Then find the approximation using 3.14 for π. $C = \pi \left(\boxed{}\ \text{cm}\right) = \boxed{}\ \pi\ \text{cm}$ Use 3.14 for π. Answer:	

Example	Notes
5. The earth's equator forms a circle. Estimate the number of miles a ship would have to travel if it went around the earth at the equator. Use 7900 miles as an approximation of the earth's diameter. Use 3.14 for π. Answer:	

Helpful Hints

Exact answers for circumference are left in terms of π.

Approximations for circumference use $\dfrac{22}{7}$ or 3.14 for the value of π.

Concept Check

1. What are the two formulas for finding the circumference of a circle?

Practice

Find the circumference of the following circles. Find the exact answer in terms of π. Then find the approximation using 3.14 for π.

2. A circle with a radius of 6 meters

3. A circle with a diameter of 7 inches

4. The bottom of an empty farm silo forms a circle. The diameter of the silo is 18 feet.

5. The lid to a trash barrel is a circle of radius 1.5 feet.

Name: _____   Date: _____
Instructor: _____   Section: _____

Introduction to Geometry
Topic 8.9 Finding Area – Circles

Vocabulary
diameter   •   area   •   radius   •   circumference

1. The _____ of a figure is the measure of the surface inside the figure.

Step-by-Step Video Notes
Watch the Step-by-Step Video lesson and complete the examples below.

Example	Notes
1. Find the area of a circle with a radius of 6 feet. Use the approximate value of 3.14 for π.	

Use the formula $A = \pi r^2$, where A is the area and r is the radius. Enter the radius.

$A = (3.14)(\square \text{ feet})^2$

Square the radius. $A = (3.14)(\square \text{ square feet})$

Answer:

Example	Notes
2. Find the area of a circle with a diameter of 6 meters. Use the approximate value of 3.14 for π. Find the radius using $r = \frac{1}{2}d$ Next use the formula $A = \pi r^2$. Answer:	
3. Find the area of a circle with a diameter of 7 feet. Use the approximate value of $\frac{22}{7}$ for π. Answer:	

Helpful Hints

Remember that area is measured in square units.

If the diameter is given when finding the area of a circle, it must be divided by 2 to get the radius.

Concept Check
1. What is the formula for finding the area of a circle?

Practice
Find the area of the following circles. Use the approximate value of 3.14 for π.
2. A circle with a radius of 10 feet
4. A circle with a diameter of 3.5 cm

3. A circle with a radius of 2.5 meters
5. A circle with a diameter of 10 mi

More Geometry
Topic 9.1 Volume – Definitions and Units

Vocabulary
Volume   •   cube   •   box   •   area

1. A _____ is a rectangular box in which every side is a square.

2. The _____ of a three-dimensional figure is the amount of space inside the figure.

Step-by-Step Video Notes
Watch the Step-by-Step Video lesson and complete the examples below.

Example	Notes
1. Find the volume of a box with a length of 4 inches, a width of 3 inches, and a height of 4 inches. How many cubes will it take to fill the bottom layer? $4 \times 3 = \boxed{}$ How many layers will fill the box? $\boxed{}$ Multiply the number of layers by the number of cubes in each layer. Answer in cubic units. Answer:	

Example	Notes

2. Find the volume of a cube that measures 4 cm on each side.
 	
  4 cm   4 cm
 	
          4 cm
 	
How many 1 cm cubes will it take to fill the bottom layer? ☐
 	
Answer:	

3. Find the volume of a box with length 5 m, width 1 m, and height 3 m.
 	
          1 m
3 m
      5 m
 	
Answer:	

Helpful Hints

Volume involves working with three dimensions; length, width, and height.

Volume is measured in cubic units. You can write out the cubic units for the label (cubic meters), or use the exponent and the abbreviation (m^3).

Concept Check
1. How does measuring area help to find volume for a cube or a box?

Practice

Find the volume of a cube with the given side measure.

2. 6 in.

3. 9 feet

Find the volume of a rectangular box with the given dimensions.

4. length 4 ft, width 2 ft, height 7 ft

5. length of 3 m, width of 6 m, height of 5 m

More Geometry
Topic 9.2 Finding Volume

Vocabulary
cylinder   •   sphere   •   pi   •   formula   •   radius

1. Using a _____ is a faster and more practical way to find the volume of a three-dimensional object than by counting unit cubes.

Step-by-Step Video Notes
Watch the Step-by-Step Video lesson and complete the examples below.

Example	Notes
1. Use the formula to find the volume of a rectangular box with a length of 12 inches, a width of 10 inches, and a height of 5 inches. $V = \text{lwh} = 12 \times 10 \times 5 = \boxed{}$ Answer:	
2. Use the formula to find the volume of a cube that measures 2.5 km on each side. Round to the nearest hundredth. $V = s^3 = (\boxed{})^3$ Answer:	

Example	Notes
3. Use $V = \pi r^2 h$ and the approximate value of 3.14 for π to find the volume of a cylinder with radius of 3 inches and a height of 5 inches. Answer:	
4. Use the formula and the approximate value of 3.14 for π to find the volume of a sphere with a radius of 6 mm. Answer:	

Helpful Hints

Use the formulas $V = lwh$, $V = s^3$, $V = \pi r^2 h$, and $V = \dfrac{4\pi r^3}{3}$ to find the volumes of rectangular boxes, cubes, cylinders, and spheres, respectively.

Concept Check
1. How does using a formula make it simpler to find the volume of an object?

Practice
Find the volume of the figure with the given dimensions.
2. a cube with a side length 8 ft

Find the volume of the figure with the given dimensions. Use 3.14 for π.
4. a cylinder with a radius of 8 in. and a height of 9 in.

3. a rectangular box with length 8 cm, width 12 cm, height 11 cm

5. a sphere with a radius of 1 meter

More Geometry
Topic 9.3 Square Roots

Vocabulary
perfect square   •   square root   •   area of a square   •   radical

1. If $a^2 = b$, then a is the _____ of b.

2. The symbol which denotes square root, $\sqrt{}$, is called the _____ sign.

Step-by-Step Video Notes
Watch the Step-by-Step Video lesson and complete the examples below.

Example	Notes
1. Find the square root of 36. When you multiply two identical factors to result in another number, the square root is one of those identical factors. The square root of 36 is $\square$, because $\square^2 = 36$. Answer:	
2. Simplify $\sqrt{64}$. Find two identical factors whose product is 64. The square root of 64 is $\square$, because $\square^2 = 64$. Answer:	
3. Simplify $\sqrt{400}$. Find two identical factors whose product is 400. Answer:	

Example	Notes
4. The area of a square is 49 cm^2. What is the length of each side? *s* [square] *s* The area of a square is found by multiplying the side length by itself. Since $\square \times \square = 49$, $\sqrt{49} = \square$. Answer:	
6. Approximate $\sqrt{75}$ by finding the two consecutive whole numbers the square root lies between. The perfect square just less than 75 is 64, and $\sqrt{64} = \square$. The perfect square just greater than 75 is $\square$, and $\sqrt{\square} = \square$. Therefore, $\sqrt{75}$ is between $\square$ and $\square$. Answer:	

Helpful Hints

A perfect square is a number that has a whole number square root. If a whole number is not a perfect square, use a calculator, or approximate the square root by finding two perfect squares that whole number lies between.

Concept Check

1. Can you find the perimeter of a square if you know its area?

Practice

Find the square root of the given number.

2. 36

3. 144

Use a calculator to approximate each square root to the nearest hundredth.

4. $\sqrt{154}$

5. $\sqrt{55}$

More Geometry
Topic 9.4 The Pythagorean Theorem

Vocabulary
Pythagorean Theorem   •   hypotenuse   •   leg   •   diagonal

1. The longest side of a right triangle, opposite the right angle, is the _____.

2. The _____ states that in a right triangle, the sum of the squares of the legs is equal to the square of the hypotenuse, or $\text{leg}^2 + \text{leg}^2 = \text{hypotenuse}^2$.

Step-by-Step Video Notes
Watch the Step-by-Step Video lesson and complete the examples below.

Example	Notes
1. Find the length of the hypotenuse of a right triangle with legs that measure 3 cm and 4 cm. To find the hypotenuse, use the formula $\text{hypotenuse} = \sqrt{\text{leg}^2 + \text{leg}^2}$. $\text{hypotenuse} = \sqrt{3^2 + 4^2} = \sqrt{9 + \square} = \sqrt{\square} = \square$ Answer:	
2. One leg of a right triangle measures 9 inches, and the hypotenuse measures 15 inches. Find the length of the other leg. To find the length of the missing leg, use the formula $\text{leg} = \sqrt{\text{hypotenuse}^2 - \text{leg}^2}$. Substitute for the hypotenuse and leg. $\text{leg} = \sqrt{15^2 - \square^2} = \sqrt{\square}$ Answer:	

Example	Notes
3. Find the missing side of the right triangle. Round to the nearest tenth.	

    Identify the formula to use. Substitute the given values into the formula. Find the missing side.

    Answer:

---

4. A standard computer monitor has a length of 20 inches and a width of 15 inches. What is the measure of the diagonal of the monitor?

    The monitor is a rectangle. The diagonal divides the rectangle into 2 identical right triangles. The length and width are the legs, and the diagonal is the hypotenuse. Identify the formula to use.

    Answer:

Helpful Hints

The Pythagorean Theorem works with right triangles or the many real-life situations where a right triangle can be applied to help find a distance. Be sure to identify whether the side of the triangle you are trying to find is a leg or the hypotenuse, and use the appropriate formula. Quite often, the Pythagorean Theorem is stated as $a^2 + b^2 = c^2$.

Concept Check

1. How do you find the length of the diagonal of a rectangle?

Practice

Find the hypotenuse of a right triangle with the given length of the legs.

2. Leg = 5 ft Leg = 12 ft

3. Leg = 16 m Leg = 12 m

Find the length of the missing leg of a right triangle given the hypotenuse and a leg.

4. Leg = 24 in. Hypotenuse = 25 in.

5. Leg = 24 cm Hypotenuse = 26 cm

More Geometry
Topic 9.5 Similar Figures

Vocabulary
similar figures   •   proportion   •   corresponding sides   •   ratio

1. _____ have the same shape and corresponding angles with the same measure.

Step-by-Step Video Notes
Watch the Step-by-Step Video lesson and complete the examples below.

Example	**Notes**
1. Identify the corresponding sides and set up proportions to determine if the figures are similar.	

The left and right sides of the smaller triangle measure 3.5 cm. The corresponding sides of the larger triangle measure 7 cm. The ratio is $\frac{3.5}{7}$.

Compare the bottom side of the smaller triangle to the bottom side of the larger triangle.

Is $\frac{3.5}{7} = \frac{\square}{\square}$?

2. Identify the corresponding sides and set up proportions to determine if the figures are similar.

Both figures are right triangles. $\frac{6}{6} = \frac{\square}{\square} = \frac{\square}{\square}$,

so the figures _____ similar.

Example	Notes
4 & 5. Determine if the figures in each pair are similar. 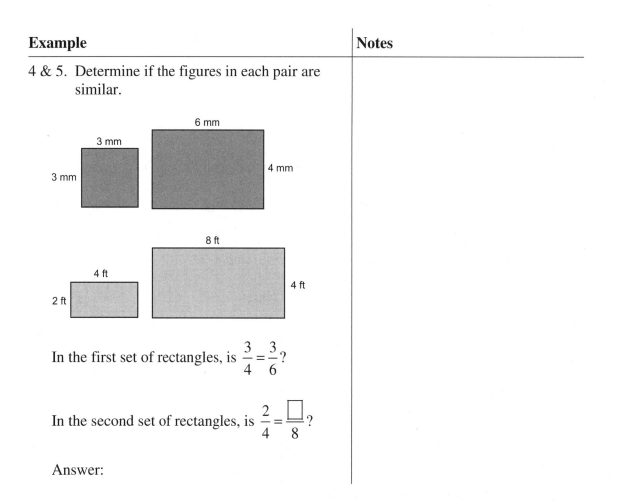 In the first set of rectangles, is $\dfrac{3}{4} = \dfrac{3}{6}$? In the second set of rectangles, is $\dfrac{2}{4} = \dfrac{\square}{8}$? Answer:	

Helpful Hints

The concept of similar figures applies to many everyday applications such as photography, art, scale models, and maps. If figures are similar, they are proportional, or "to scale."

If the corresponding sides of two triangles are proportional, the triangles are similar. Likewise, two rectangles are similar if their corresponding sides are proportional.

Concept Check

1. Can you think of geometric shapes other than squares that are always similar?

Practice

Determine if the figures are similar.

2. Triangle A with sides 4 m, 5 m, 6m
    Triangle B with sides 8 m, 12 m, 10 m

3. Triangle C with sides 9 ft, 9 ft, 7 ft
    Equilateral Triangle D with sides 9 ft

Can the following pairs of figures be similar?

4. A rectangle and a trapezoid

5. A very small square measured in mm and a very large square measured in km

Name: _____   Date: _____

Instructor: _____   Section: _____

More Geometry
Topic 9.6 Finding Unknown Lengths

Vocabulary

corresponding sides   •   proportion   •   proportional   •   unknown

1. Corresponding sides of similar figures are _____.

Step-by-Step Video Notes
Watch the Step-by-Step Video lesson and complete the examples below.

Example	Notes
1 & 2. Write a proportion for the corresponding sides of the similar figures.	

The proportion for the corresponding sides of

the first pair of triangles is $\dfrac{2.5}{5} = \dfrac{\square}{12} = \dfrac{6.5}{\square}$.

Each ratio in the proportion is equal to $\dfrac{1}{\square}$.

In the second pair, the figures have exactly the same shape, therefore they are _____.

Answer:

Example	Notes

3. Find the unknown side length in the similar figures.

The proportion is $\dfrac{\square}{x} = \dfrac{8}{\square}$.

Answer:

---

4. Find the unknown side length in the similar figures.

The proportion is $\dfrac{4}{\square} = \dfrac{x}{\square}$.

Answer:

Helpful Hints

If figures are similar, the ratios of the corresponding sides will be equal. Line up the corresponding sides of similar figures in a proportion to find unknown lengths.

Concept Check

1. Is a 30 in. wide by 36 in. tall poster similar to a 4 in. by 6 in. photo?

Practice

Solve for x.

2. A 6 in. by 9 in. rectangle is similar to a 24 in. by x in. rectangle

3. A triangle with sides 6 m, 10 m, 8 m is similar to a triangle with sides 9 m, 15 m, and x m.

Find the unknown length.

4. A 5 cm by 12 cm rectangle is similar to a 15 mm by ☐ mm rectangle.

5. $\dfrac{8 \text{ ft}}{12 \text{ ft}} = \dfrac{10 \text{ ft}}{\boxed{}}$

Name: _____ Date: _____

Instructor: _____ Section: _____

Statistics
Topic 10.1 Bar Graphs and Histograms

Vocabulary
bar graph • double bar graph • measurement • frequency distribution table • histogram

1. A _____ is a special type of bar graph where the width of the bar represents an interval, or range of numbers.

Step-by-Step Video Notes
Watch the Step-by-Step Video lesson and complete the examples below.

Example	Notes
2.1 Use the bar graph below to answer the question. How many Country Music Association Awards did Loretta Lynn win?	

Loretta Lynn
Country Music Association Awards

The height of each bar represents the number in each category. Enter these numbers from the graph.

Number of awards = $4 + \boxed{} + \boxed{}$

Answer:

2.2 Use the bar graph in example #1 to answer the question. In which category did she win the fewest awards?

Which bar is the lowest? _____

Example	Notes

5. In an Intro to Music class, letter grades are distributed based on the following scale: score of 90-100 an A, 80-89 a B, 70-79 a C, 60-69 a D, and below 60 is an F. Enter the tally and frequency in the table below for a class with scores of 78, 69, 82, 95, 92, 80, 47, 89, 81, and 99.

Intro To Music Grade Distribution		
Grade Intervals	Tally	Frequency
90 – 100		
80 – 89		
70 – 79		
60 – 69		
Below 60		

Helpful Hints

A bar graph can be used to display data over time, to compare amounts, or show how often a particular amount will occur.

The height of the rectangular bar indicates the number in each category in a bar graph.

Concept Check

1. Name two similarities and one difference between a bar graph and a histogram.

Practice

Use the bar graph below to answer the following.

2. How many softball team members voted?

Softball Team Captain Vote

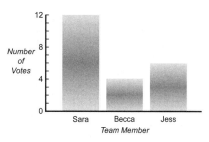

3. Which player received the fewest votes?

Fill in the frequency distribution table for the given data below.

4. The Human Resources Manager at a company tracks employee sick days in the categories of 0-2, 3-5, 6-8, 9-11, and 12 or more. The art department employees used sick days of 4, 6, 2, 3, 9, 0, 1, 0, 3, 2, 4, 7, 0, 1, and 15.

Employee Sick Days Distribution		
# of days used	Tally	Frequency
0 – 2		
3 – 5		
6 – 8		
9 – 11		
12 or more		

Name: _____   Date: _____

Instructor: _____   Section: _____

Statistics
Topic 10.2 Line Graphs

Vocabulary
double line graphs   •   line graphs   •   data points   •   histograms

1. _____ can display data, show trends or patterns in data over time, show how often a particular data value occurs, compare two or more types of data, and use data points that are connected with straight line segments.

Step-by-Step Video Notes
Watch the Step-by-Step Video lesson and complete the examples below.

Example	Notes
1.1 Use the line graph below to answer the question. Has the average life expectancy of humans increased or decreased since 1920? **Average Life Expectancy of Humans from 1920 to 2000** *Years of Life* (vertical axis: 60, 70, 80) *Decades in the 1900s* (horizontal axis: 1920 1940 1960 1980 2000) Answer:	
1.2 Use the line graph in example #1 to answer the question. During which 20-year interval did the life expectancy increase the most? Answer:	

Example	Notes
3.1 Use the double line graph below to answer the question. Which countries had the same life expectancy in for 2000 and 1998?	

Life Expectancy by Country
1998 and 2000

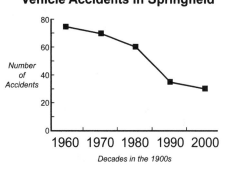

● 1998   ■ 2000

Answer:

Helpful Hints

A line graph is usually used to show trends or patterns of data over time, while a double line graph is used to compare more than one set of data on a graph.

Concept Check

1. State at least two differences between a bar graph and a line graph.

Practice

Use the bar graph below to answer the following.

2. Have the vehicle accidents in Springfield increased or decreased since 1960?

Vehicle Accidents in Springfield

Use the double bar graph below to answer the following.

4. During which decade were there more vehicle accidents in Newton than in Springfield?

Vehicle Accidents

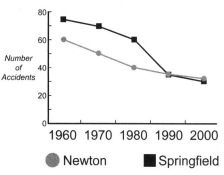

● Newton   ■ Springfield

3. During which decade was the decrease in accidents the most?

Statistics
Topic 10.3 Circle Graphs

Vocabulary
line graph   •   circle graph   •   amounts   •   percents

1. A _____, also called a pie chart, is drawn to show actual amounts or percents
   and is used to show how one set of data is divided.

Step-by-Step Video Notes
Watch the Step-by-Step Video lesson and complete the examples below.

Example	Notes
1.1 Use the circle graph below to answer the question. What does Nikeshia plan to spend the most money on? **Average Costs of School** *Books* $400 *Tuition* $1200 *Housing* $800 *Food* $800 $400 *Other* Answer:	
1.2 Use the circle graph in example #1 to answer the question. What total amount does Nikeshia plan to spend on food and housing? Find the sum of the amount from the food category and the housing category. Answer:	

Example	Notes
3.1 Use the circle graph below to answer the question. Which assignment type counts the most toward the student's final grade? **Breakdown of Writing Class Grade Components** 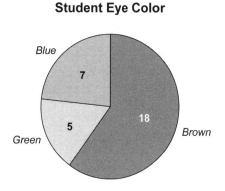 Answer:	

Helpful Hints

Circle graphs are most often used to display relationships that compare parts to a whole.

In circle graphs the sum of all of the parts must equal the total amount or 100%.

Concept Check

1. What is different about a circle graph when compared to a bar graph or line graph?

Practice

Use the circle graphs to answer the following questions.

2. What eye color is the least frequent in this class of students?

4. What percentage of students in this class has blue eyes?

Student Eye Color

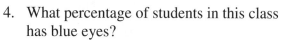

Student Eye Color

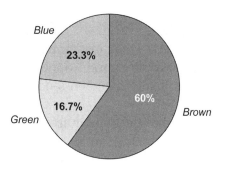

3. How many students are in this class?

Name: _____   Date: _____

Instructor: _____   Section: _____

Statistics
Topic 10.4 Mean

Vocabulary

measures of central tendency   •   statistics   •   mean   •   data

1. _____ is information about a group or topic, often consisting of a set of numbers.

2. The _____ is the sum of the values in the list of numbers divided by the number of values.

Step-by-Step Video Notes
Watch the Step-by-Step Video lesson and complete the examples below.

Example	Notes
1. Calculate the mean for the following list of numbers. 3, 5, 7, 2, 11, 6, 8, 2, 10 Find the sum of all of the values in the list. $3 + 5 + \boxed{} + 2 + 11 + 6 + 8 + \boxed{} + 10 = \boxed{}$ There are 9 values in the list. Divide the sum by the number of values in the list. $\boxed{} \div 9 = \boxed{}$ Answer:	
2. Calculate the mean of the following list of test scores. 93, 86, 95, 98, 82 Find the sum. $\boxed{}$ There are $\boxed{}$ values in the list. Divide the sum by the number of values in the list. Answer:	

Example	Notes
3. Eli saved a portion of his allowance for 6 weeks. He saves $14 the first week, $12 the second week, $6 the third week, and $4, $10, and $2 for the remaining weeks. On average, how much did he save each week? Find the average, which is the same as the mean, saved over the six weeks.	

Helpful Hints

Mean is a measure of central tendency because it represents an entire list of numbers as well as the "center" of the data.

The mean is often referred to as the average.

Concept Check

1. Describe how to find the mean of a list of numbers.

Practice

Find the mean of each list of numbers.

2. 10, 8, 4, 7, 5, 8, 10, 4

3. 55, 42, 33, 61, 30

4. The five players on Jamison's basketball team collected donations for new team uniforms. Jamison collected $18. His teammates collected $12, $34, $26 and $10. On average, how much did each player collect?

Name: _____   Date: _____

Instructor: _____   Section: _____

Statistics
Topic 10.5 Median

Vocabulary

median   •   statistics   •   mean   •   data

1. _____ is the study and use of data.

2. The _____ of a set of ordered numbers is the middle number.

Step-by-Step Video Notes
Watch the Step-by-Step Video lesson and complete the examples below.

Example	Notes
1. Find the median of the following list of numbers. 3, 5, 7, 2, 11, 6, 8, 2, 10 Write the numbers in order from smallest to largest. 2, ☐ , 3, 5, ☐ , ☐ , 8, 10, ☐ Does the list contain an odd number or even number of values? Odd; there are 9 values. Mark off one number from the beginning of the list and from the end of the list. 2̸ , 2, 3, 5, ☐ , ☐ , 8, 10, 1̸1̸ Repeat this step. 2̸ , 2̸ , 3, 5, ☐ , ☐ , 8, 1̸0̸ , 1̸1̸ Repeat this step until the middle number is reached. Answer:	

Example	Notes
2. Find the median of the following list of numbers. 44, 32, 31, 56, 77, 65 Order the numbers. Since there are an even number of values, the median is the average of the two middle numbers. Find this average. $44 + \boxed{} = \boxed{}$ $\boxed{} \div 2 = \boxed{}$ Answer:	
3. A poll of nine students shows the numbers of Facebook friends they have are 395, 486, 166, 430, 172, 159, 723, 582, and 319. Find the median number of Facebook friends for these nine students. Answer:	

Helpful Hints

Remember that the values must be ordered before finding the median of a list of numbers.

Concept Check

1. How do you find the median of a list that has an even number of values? How does this differ from the process of finding the median of list that has an odd number of values?

Practice

Find the median of each list of numbers.

2. 10, 8, 4, 7, 5, 8, 10

4. Sarah's math test scores for the first term were 95, 92, 81, 98, 84, 86, and 94. Find the median of Sarah's math test scores.

3. 55, 43, 33, 61, 30, 41

Statistics
10.6 Mode

Vocabulary

median   •   statistics   •   data   •   mode

1. The _____ of a list of numbers is the number that occurs most often in the list.

Step-by-Step Video Notes
Watch the Step-by-Step Video lesson and complete the examples below.

Example	Notes
1. Find the mode(s) of the following list of numbers. 3, 5, 7, 2, 11, 6, 8, 2, 10 Write the numbers in order from smallest to largest. 2, 2, 3, ☐, 6, ☐, ☐, 10, 11 Choose the number(s) that occur the most often. Answer:	
2. Find the mode(s) of the following list of numbers. 44, 32, 31, 56, 77, 65, 65, 44, 65 Order the numbers. 31, ☐, 44, ☐, ☐, ☐, ☐, 65, 77 Answer:	
3. Find the mode(s) of the following list of numbers. 325, 612, 367, 692, 451 Note that a list of numbers can have more than one mode or no mode. Answer:	

Example	Notes
5. According to the National Weather Service, the high temperatures (in °F) for the first two weeks of June 2010 were recorded as 105, 106, 109, 112, 116, 119, 119, 115, 111, 107, 106, 103, 104, and 110. Find the mode(s) of the high temperatures for this period. Answer:	

Helpful Hints

It is not a required step to order the numbers when finding the mode, but it can be helpful.

The mode is not just any number that occurs more than once, it is the number that occurs more often than the rest.

Remember that there can be one mode, more than one mode, or no mode.

Concept Check
1. How is the mode different than the other two measures of central tendency, the mean and the median?

Practice
Find the mode(s) of each list of numbers.

2. 65, 35, 40, 55, 40, 30, 70

3. 10, 8, 4, 7, 5, 8, 10

4. The students in a kindergarten class each have a box for storing crayons and pencils. The number of crayons in the boxes is 8, 6, 7, 6, 2, 5, 7, 4, 3, and 8. Find the mode(s) of the crayons in the boxes.

Real Numbers and Variables
Topic 11.1 Introduction to Real Numbers

Vocabulary

| whole numbers | • | integers | • | rational number |
| irrational number | • | real numbers | • | radical sign |

1. _____ are all the rational numbers and all the irrational numbers.

2. The decimal form of a(n) _____ is non-terminating and non-repeating.

Step-by-Step Video Notes

Watch the Step-by-Step Video lesson and complete the examples below.

Example	Notes

1. Classify each number as an integer, a rational number, an irrational number, and/or a real number. Circle all that apply.
 	
   8   integer   rational   irrational   real
 	
   $-\dfrac{1}{5}$ integer   rational   irrational   real
 	
   $\sqrt{3}$ integer rational irrational real	

2–4. Plot the following real numbers on a number line. 1 -2.75 $\dfrac{1}{2}$	

Example	Notes
5–7. Use a real number to represent each real-life situation. A temperature of 131.2 °F below zero is recorded in Antarctica. The height of a mountain is 22,645 feet above sea level. A golfer scored 5 under par in a recent tournament.	

Helpful Hints

Rational numbers are any numbers that can be written as fractions in the form $\frac{a}{b}$, where a and b are integers. The decimal equivalents of rational numbers repeat or terminate.

Irrational numbers, such as π or $\sqrt{2}$, are non-terminating, non-repeating decimals.

Positive numbers are to the right of 0 on a number line. Negative numbers are to the left of 0 on a number line.

Concept Check

1. You would plot $-\frac{7}{2}$ between what two integers on a number line?

Practice

Classify each number as rational or irrational.

2. 0.3333...

3. $\sqrt{16}$

Use a real number to represent the situation.

4. New Orleans is 64 feet below sea level.

5. Kelsey deposits $125 into her savings account.

Real Numbers and Variables
Topic 11.2 Graphing Rational Numbers Using a Number Line

Vocabulary
number line   •   inequality   •   inequality symbols

1. A(n) _____ is a statement that shows the relationship between any two real numbers that are not equal.

2. The _____ ">" and "<" are used to represent the phrases " is greater than" and "is less than."

Step-by-Step Video Notes
Watch the Step-by-Step Video lesson and complete the examples below.

Example	Notes
1 & 2. Plot the following real numbers on a number line. 2 ⊢—⊢—⊢—⊢—⊢—⊢—⊢→ −3 −2 −1 0 1 2 3 −1.5 ⊢—⊢—⊢—⊢—⊢—⊢—⊢→ −3 −2 −1 0 1 2 3	
3 & 4. Write the following statements using inequality symbols. 2 is less than 7 8 is greater than 5	

Example	Notes
5–7. Plot the given numbers on a number line, and then replace the question mark with the appropriate symbol, > or <. 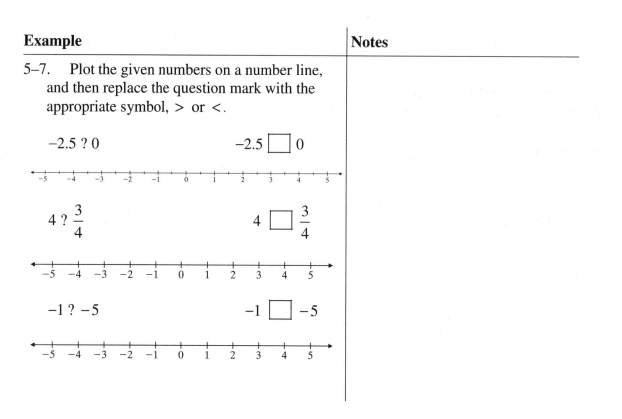	

Helpful Hints

When comparing numbers, one number is greater than another number if it is to the right of that number on the number line. One number is less than another number if it is to the left of that number on the number line.

Remember that the inequality symbols ">" and "<" point toward the lesser value.

Concept Check

1. How can you tell which is the greater of two numbers plotted on a number line?

Practice

Plot the real numbers on a number line.

2. $2\frac{2}{3}$

3. −0.5

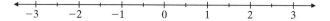

Write statements using inequality symbols.

4. 58 is less than 64

5. −7 is greater than −10

Real Numbers and Variables
Topic 11.3 Translating Phrases into Algebraic Inequalities

Vocabulary
variable   •   inequality phrase   •   inequality symbols

1. A(n) _____ is a letter or symbol that is used to represent an unknown quantity.

Step-by-Step Video Notes
Watch the Step-by-Step Video lesson and complete the examples below.

Example	Notes
1. Translate the phrase into an algebraic inequality. A police officer claimed that a car was traveling at a speed more than 85 miles per hour. (Use the variable s for speed.) Determine the phrase to be translated. Replace the unknown quantity with the given variable. Replace the inequality phrase with the correct inequality symbol. Answer:	
3. Translate the phrase into an algebraic inequality. The owner of a trucking company said that the payload of a truck must be no more than 4500 pounds. (Use the variable p for payload.) Determine the phrase to be translated. Replace the unknown quantity with the given variable. Answer:	

Example	Notes
4. Translate the phrase into an algebraic inequality.	

Carlos must be at least 16 years old in order to get his driving license. (Use the variable a for age.)

Determine the phrase to be translated.

Replace the unknown quantity with the given variable.

Answer:

Helpful Hints

When translating phrases into algebraic inequalities look for key words or phrases to determine which inequality symbol is most appropriate to use.

The inequality symbol "$\leq$" can also mean "at most" or "no more than." The inequality symbol "$\geq$" can also mean "at least" or "no less than."

Concept Check

1. Is $-5 \leq -5$? Is $\dfrac{7}{4} \geq 1\dfrac{3}{4}$?

Practice

Translate each phrase into an algebraic inequality.

2. According to the building inspector, the elevator can hold at most 12 people. (Use the variable n for number of people.)

3. A truck must be no taller than 9.5 feet to go through a certain tunnel. (Use the variable h for the height of the truck.)

4. You must be at least 54 inches tall to go on the roller coaster. (Use the variable i for the number of inches.)

Real Numbers and Variables
Topic 11.4 Finding the Absolute Value of a Real Number

Vocabulary

distance • number line • opposites • absolute value

1. The _____ of a number is the distance between that number and zero on a number line.

Step-by-Step Video Notes
Watch the Step-by-Step Video lesson and complete the examples below.

Example	Notes
1 & 2. Use a number line to find the absolute value of the following numbers. 3 -1.5	
3. Write the expression for the absolute value of $-\dfrac{3}{8}$. _____	
4–6. Find the absolute value of the following numbers. $\left\|-3.68\right\| = \boxed{}$ $\left\|3.68\right\| = \boxed{}$ $\left\|0\right\| = \boxed{}$	

Example	**Notes**

7–9.   Find the absolute value of the following numbers.

$$|-8.333...| = \boxed{}$$

$$\left|5\frac{9}{10}\right| = \boxed{}$$

$$\left|-\frac{2}{7}\right| = \boxed{}$$

Helpful Hints

The absolute value of a number can be thought of as the numerical part of the number, without regard to its sign.

Absolute value is the distance from zero on a number line. Since it represents a distance, absolute value can never be negative.

Concept Check

1.   How many numbers have an absolute value of 6.5? of $-\frac{3}{4}$?

Practice

Write the expression for the absolute value of the numbers given.

2.   −.8888...

Find the absolute value of the following numbers.

4. $|-24|$

3. $\frac{1}{4}$

5. $\left|-6\frac{7}{8}\right|$

Adding and Subtracting Real Numbers
Topic 12.1 Adding Real Numbers with the Same Sign

Vocabulary
absolute values • positive numbers • negative numbers • same sign

1. When adding add two numbers of the _____, add the absolute values, or numerical parts, of the numbers and give the answer the same sign as the numbers being added.

Step-by-Step Video Notes
Watch the Step-by-Step Video lesson and complete the examples below.

Example	Notes
1 & 2. Use chips to add the following. $4+2$ $\oplus \oplus \oplus \oplus + \oplus \oplus = \oplus \oplus \oplus \oplus \oplus \oplus$ Answer: $-2+(-3)$ $\odot \odot + \odot \odot \odot = \odot \odot \odot \odot \odot$ Answer:	
4. Add $-3+(-5)$. Add the numerical parts. Give the answer the sign of the numbers being added. Answer:	

Example	Notes
5. Add $-\dfrac{2}{3}+\left(-\dfrac{1}{7}\right)$. Add the numerical parts. Find a common denominator. Answer:	
6. Add $-8.1+(-2.75)+(-5.03)$. Add from left to right. Answer:	

Helpful Hints

To add numbers with the same sign, add the numerical parts of the numbers and give the answer the same sign as the numbers being added.

A number written without a sign is assumed to be positive.

Concept Check

1. What will be the sign of the sum of $-\dfrac{3}{5}+\left(-\dfrac{4}{9}\right)$?

Practice
Add.

2. $6.27+12.8$

3. $-\dfrac{4}{5}+\left(-\dfrac{7}{10}\right)$

4. $-8+(-8)$

5. $-8.43+(-0.57)$

Adding and Subtracting Real Numbers
Topic 12.2 Adding Real Numbers with Different Signs

Vocabulary
different signs   •   numerical part   •   common denominator

1. When adding two numbers with _____, subtract the absolute values, or numerical parts, of the numbers and give the answer the same sign as the larger numerical part.

Step-by-Step Video Notes
Watch the Step-by-Step Video lesson and complete the examples below.

Example	Notes
1 & 2. Use chips to add the following.	
$6+(-4)$	
$\oplus\,\oplus\,\oplus\,\oplus\,\oplus\,\oplus$ $+\odot\,\odot\,\odot\,\odot$ ___	
Answer:	
$-7+3$	
$\odot\,\odot\,\odot\,\odot\,\odot\,\odot\,\odot$ $+\oplus\,\oplus\,\oplus$ _____	
Answer:	
3. Add $-6+9$.	
Subtract the numerical parts.	
Give the answer the same sign as the larger "numerical part."	
Answer:	

Example	Notes
4. Add $\dfrac{3}{7}+\left(-\dfrac{5}{7}\right)$.	
Answer:	

5. Add $3.7+(-10.5)$.

Answer:

Helpful Hints
Remember that the absolute value of a number is the distance between that number and zero on a number line.

Try to determine the sign of the answer before calculating. You're less likely to forget to give the answer the correct sign.

Concept Check
1. What will be the sign of the sum of $5+\left(-3\dfrac{2}{9}\right)$?

Practice
Add.

2. $6+(-13)$

3. $\dfrac{4}{7}+\left(-\dfrac{2}{7}\right)$

4. $-7.7+8.7$

5. $-\dfrac{7}{8}+\dfrac{2}{3}$

Adding and Subtracting Real Numbers
Topic 12.3 Finding the Opposite of a Real Number

Vocabulary
opposite numbers   •   additive inverses   •   absolute value

1. Two numbers that differ only in sign are _____. They are the same distance from zero on a number line but in opposite directions.

2. Two numbers are _____, or opposites, if they add to equal zero.

Step-by-Step Video Notes
Watch the Step-by-Step Video lesson and complete the examples below.

Example	Notes
1. Find the opposite of the situation. A temperature increase of 8.5 ° F. The opposite of increase is _____. The opposite of the situation is a _____ of 8.5 ° F.	
3–5. Find the opposite of the following numbers. 7 -3.5 $-\dfrac{5}{7}$ To find the opposite of a number, change the sign of the number. The opposite of 7 is ☐ . The opposite of −3.5 is ☐ . The opposite of $-\dfrac{5}{7}$ is ☐ .	

Example	Notes
6 & 7. Find the opposite of the following absolute values. $\lvert 5 \rvert$ $\quad\quad\quad$ $\lvert -6.78 \rvert$ $-\lvert 5 \rvert = \square$ $-\lvert -6.78 \rvert = \boxed{}$	
8. Find the additive inverse, or opposite, of -4. Then add the additive inverse to the number. The additive inverse of -4 is $\square$. $-4 + \left(\boxed{} \right) = \square$	

Helpful Hints

For any real number a, $-(-a) = a$, but $-\lvert -a \rvert = -a$.

For any real number, its opposite and its additive inverse are the same number.

Concept Check

1. What will be the sign of the additive inverse of $\left(-6\dfrac{7}{8} \right)$?

Practice

Find the additive inverse, or opposite, of each number.

2. -2.3

Find the opposite of the following absolute values.

4. $\lvert 4.4 \rvert$

3. $-\left(-\dfrac{4}{5} \right)$

5. $\left\lvert -\dfrac{2}{3} \right\rvert$

Adding and Subtracting Real Numbers
Topic 12.4 Subtracting Real Numbers

Vocabulary
difference   •   add the opposite   •   sum

1. To subtract real numbers, _____ of the second number to the first. This
   sometimes is referred to this as leave, change, change.

Step-by-Step Video Notes
Watch the Step-by-Step Video lesson and complete the examples below.

Example	**Notes**
1. Subtract $10 - 40$. Leave the first number alone. Change the minus sign to a plus sign. Change the sign of the number being subtracted. Add the two numbers using the rules of addition. Answer:	
3. Subtract $-5 - (-9)$. Answer:	

Example	Notes
4. Subtract $-6.18 - 2.34$. Answer:	
5. The temperature in a city is $-8\,°\text{F}$ and the temperature of a neighboring town is $-15\,°\text{F}$. Find the difference in temperature between the two cities. Write the difference as a subtraction problem. Answer:	

Helpful Hints
Subtracting a number is the same as adding its opposite.

Any subtraction problem with real numbers can be changed to an addition problem by using the methods "Add the Opposite" or "Leave, Change, Change."

Concept Check
1. What will always be the sign of the difference when you subtract a positive number from a negative number?
2. What will always be the sign of the difference when you subtract a negative number from a positive number?

Practice

Subtract. Use "Add the Opposite."

3. $-2 - (-9)$

4. $\dfrac{4}{5} - \left(-\dfrac{3}{5}\right)$

Subtract. Use "Leave, Change, Change."

5. $-2.7 - 5.4$

6. $5\dfrac{1}{3} - 6\dfrac{1}{2}$

Adding and Subtracting Real Numbers
Topic 12.5 Addition Properties of Real Numbers

Vocabulary
Commutative Property of Addition   •   Addition Property of Zero   •   Associative
Property of Addition   •   Additive Inverse Property   •   Distributive Property

1. The _____ states that changing the order of the numbers being added does not change the sum.

2. The _____ states that changing the grouping when adding numbers does not change the sum.

3. In symbols, the _____ states that for any real number a, $a + (-a) = 0$ and $-a + a = 0$.

Step-by-Step Video Notes
Watch the Step-by-Step Video lesson and complete the examples below.

Example	Notes
1–3. Determine which property of addition is exhibited by each equation. The properties are Commutative, Associative, Addition Property of Zero, and Additive Inverse Property.	
$0 + (-7) = -7$ _____	
$(-2 + 3) + 5 = 5 + (-2 + 3)$ _____	
$10 + (-10) = 0$ _____	

Example	Notes
4 & 5. Determine which property of addition is exhibited by each equation. The properties are Commutative, Associative, Addition Property of Zero, and Additive Inverse Property. $(-2+6)+0.3=-2+(6+0.3)$ _____ $-\dfrac{2}{3}+\dfrac{1}{3}=\dfrac{1}{3}+\left(-\dfrac{2}{3}\right)$ _____	

Helpful Hints

Use the properties of addition in cases where a property may make it easier for you to simplify an expression.

Concept Check

1. Which property of addition will make it easier for you to simplify the expression $6.2+(3.8+(-7.2))$? Explain how you would use this property in this case.

Practice

Match each description with the addition property that it describes.

2. This addition property deals with changing the order of the numbers being added.

3. This addition property deals with adding a number to its opposite.

Write an equivalent expression using the Associative Property of Addition.

4. $5.2+(-5.2+(-7.2))$

5. $(-8.6+5)+9.03$

Name: _____   Date: _____

Instructor: _____   Section: _____

Multiplying and Dividing with Real Numbers
Topic 13.1 Multiplying Real Numbers

Vocabulary
absolute values   •   product   •   negative factors

1. When multiplying two numbers with the same sign, the _____ will always be positive.
2. When multiplying an odd number of _____, the product is negative.

Step-by-Step Video Notes
Watch the Step-by-Step Video lesson and complete the examples below.

Example	**Notes**
1. Multiply $(7)(-4)$. Multiply the numerical parts of the numbers. $(7)(4) = \boxed{}$ Determine the sign of the answer. $(+)(-) = \boxed{}$ Answer:	
3. Multiply $(-12)(-9)$. Answer:	
4. Multiply $(-3.5)(-2)$. Answer:	

Example	Notes
5–8. Multiply.	

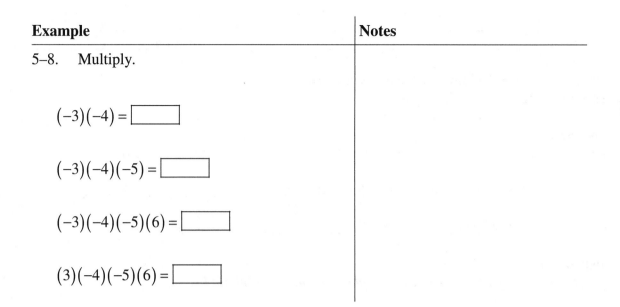

$(-3)(-4) = \boxed{}$

$(-3)(-4)(-5) = \boxed{}$

$(-3)(-4)(-5)(6) = \boxed{}$

$(3)(-4)(-5)(6) = \boxed{}$

Helpful Hints

When multiplying two numbers with the same sign, the product is positive. When multiplying two numbers with different signs, the product is negative.

When multiplying an even number of negative factors, the product is positive. When multiplying an odd number of negative factors, the product is negative.

Concept Check

1. Will the product of $-(-27)(-132)$ be positive or negative? Explain.

Practice

Multiply.

2. $(-0.8)(5)$

4. $(-2)(-5)(-3)$

3. $\left(-\dfrac{1}{4}\right)\left(-\dfrac{2}{3}\right)$

5. $(-2)(-5)(-3)(-8)$

Multiplying and Dividing with Real Numbers
Topic 13.2 Finding the Reciprocal of a Real Number

Vocabulary
integer   •   reciprocals   •   multiplicative inverse   •   invert

1. Two numbers are _____ of each other if their product is 1.

2. The reciprocal is also called the _____ of a number.

Step-by-Step Video Notes
Watch the Step-by-Step Video lesson and complete the examples below.

Example	Notes
1. Find the reciprocal of $-\dfrac{5}{7}$. The number is already written as a fraction. Invert the fraction. The sign of the reciprocal is the same as the sign of the original number. Answer:	
2. Find the reciprocal of -3. Write the number as a fraction with 1 as the denominator. Invert the fraction. Answer:	

Example	Notes
3. Find the reciprocal of 0.5 . Write the number as a fraction with 1 as the denominator. Answer:	
4. Find the reciprocal of $-4\frac{3}{5}$. Write the number as an improper fraction. Answer:	

Helpful Hints

A negative fraction can be written in 3 different but equal ways: $-\dfrac{a}{b}=\dfrac{-a}{b}=\dfrac{a}{-b}$.

To find the reciprocal of an integer or a decimal, start by putting the number over 1.

To find the reciprocal of a mixed number, first write the number as an improper fraction.

Concept Check
1. What is the sign of the reciprocal of a negative number? A positive number?

Practice
Find the reciprocal.

2. $-\dfrac{3}{8}$

4. $-5\dfrac{1}{3}$

3. 4

5. 0.4

Name: _____   Date: _____

Instructor: _____   Section: _____

Multiplying and Dividing with Real Numbers
Topic 13.3 Dividing Real Numbers

Vocabulary
division   •   quotient   •   dividend   •   divisor   •   mixed number

1. The operation of splitting a quantity or number into equal parts is_____.

2. The number you are dividing by in a division problem is called the _____.

3. When dividing numbers with different signs, the _____ will be negative.

Step-by-Step Video Notes
Watch the Step-by-Step Video lesson and complete the examples below.

Example	**Notes**
1–3. Divide. $(-21) \div (-7)$ $3.2 \div 0.8$ $\dfrac{-45}{25}$	
4. Four friends decide to start a business together. They share a start-up loan of $120,000$. If they split the amount of the loan equally between them, how much does each friend owe? The loan can be represented as a negative number. $(-120,000) \div 4 = \boxed{}$ Answer:	

Example	Notes
5 & 6. Divide. $32 \div \left(-\dfrac{8}{3}\right)$ $\left(-\dfrac{11}{6}\right) \div 2\dfrac{4}{9}$	

Helpful Hints

The fraction bar in $\dfrac{a}{b}$ is another way to express division. The expression $\dfrac{a}{b}$ is the same as $a \div b$. Follow the rules for division to simplify.

When dividing by a fraction, find the reciprocal of the divisor, and multiply. If the divisor is a mixed number, rewrite it as an improper fraction and divide as stated above.

Concept Check

1. Rewrite the division problem $\left(\dfrac{-21}{25}\right) \div (-7)$ to show the equivalent multiplication by the reciprocal of the divisor.

Practice
Divide.

2. $-56 \div (-7)$

4. $12.8 \div (-0.8)$

3. $\left(\dfrac{-12}{35}\right) \div \left(-\dfrac{6}{7}\right)$

5. $14 \div \left(-1\dfrac{3}{4}\right)$

Multiplying and Dividing with Real Numbers
Topic 13.4 Exponents and the Order of Operations

Vocabulary
base   •   exponent   •   order of operations   •   even power

1. An exponent is used as a shortcut for repeated multiplication. The _____
   is the number being multiplied.

Step-by-Step Video Notes
Watch the Step-by-Step Video lesson and complete the examples below.

Example	Notes
1. Write in exponential form. $\left(-\dfrac{5}{7}\right)\left(-\dfrac{5}{7}\right)\left(-\dfrac{5}{7}\right)\left(-\dfrac{5}{7}\right)$ Answer:	
4. Evaluate. $\left(-\dfrac{2}{5}\right)^3$ Answer:	
7 & 8. Evaluate. $\left(-5\right)^2$ $-\left(-\dfrac{1}{5}\right)^3$	

Example	Notes
9 & 10. Evaluate. $$\frac{4^3 + 2(-5)}{2^3}$$ $$(-4)^2 - 2(5)^2$$	

Helpful Hints

Any non-zero number raised to the zero power is equal to 1. Any number raised to the 1 power is equal to itself.

A negative base raised to an even power is positive. A negative base raised to an odd power is negative. Notice that $(-2)^4 = 16$, while $-2^4 = -16$.

The fraction bar acts like parentheses when evaluating an expression. Find the value of the expression in the numerator and the value of the expression in the denominator, then divide.

Concept Check

1. What is the sign of the product $(-9)(-7)^3$?

Practice

Evaluate.

2. $\left(-\dfrac{1}{4}\right)^1$

3. $(456789321)^0$

4. $\dfrac{-6 + (-5)^2}{-(-2)^3 + 3}$

5. $(-0.4)^2 + (0.3)^2 - (0.25)^1$

Multiplying and Dividing with Real Numbers
Topic 13.5 The Distributive Property

Vocabulary

mental multiplication   •   distributive property   •   order of operations

1. The _____ states that for all real numbers a, b, and c,

    $a(b+c) = ab + ac$.

Step-by-Step Video Notes

Watch the Step-by-Step Video lesson and complete the examples below.

Example	Notes
1. Multiply $9 \cdot 103$. Rewrite as an addition problem. $9 \cdot 103 = \boxed{}\left(\boxed{} + \boxed{}\right)$ Use the Distributive Property to multiply, and then simplify. $9(100) + \left(\boxed{}\right)\left(\boxed{}\right)$ Answer:	
3. Multiply $-3(-8x+3)$. Rewrite using the Distributive Property. $-3(-8x+3) = (-3)(-8x) + \left(\boxed{}\right)\left(\boxed{}\right)$ Simplify the result. $(-3)(-8x) + (-3)(3) = \boxed{}$ Answer:	

Example	Notes
4–7. Multiply. $2(-4y+7) = \boxed{}$ $-9(-4x-6) = \boxed{}$ $-(x-2) = \boxed{}$ $5(3x+2y-6z) = \boxed{}$	

Helpful Hints

Use the Distributive Property to make mental multiplication easier whenever you can change one of the factors into a sum where one of the addends is an easy number to multiply.

The Distributive Property is important and useful in simplifying algebraic expressions when you cannot perform the operations inside parentheses. To remove the parentheses, be sure to multiply by each term inside the parentheses.

Concept Check

1. What must you multiply each term in the parentheses by to simplify $-(x+y-4)$?

Practice

Rewrite using the Distributive Property.

2. $9\left(-\dfrac{1}{3}+\dfrac{2}{9}\right)$

Simplify using the Distributive Property.

4. $-6(-5x+4)$

3. $104(-8)$

5. $9(-3x+4y-7z)$

Multiplying and Dividing with Real Numbers
Topic 13.6 Multiplication Properties of Real Numbers

Vocabulary
Commutative Property   •   Zero Property   •   Associative Property
Identity Property

1. The _____ of Multiplication states that changing the order when multiplying numbers does not change the product.

2. The _____ of Multiplication states that changing the grouping when multiplying numbers does not change the product.

Step-by-Step Video Notes
Watch the Step-by-Step Video lesson and complete the examples below.

Example	**Notes**
1–4. Determine which property of multiplication is shown by each equation: the Commutative Property of Multiplication, the Associative Property of Multiplication, or the Identity Property of Multiplication. $(-17) \cdot 1 = -17$ Answer: $(-4 \cdot 2) \cdot 3 = -4(2 \cdot 3)$ Answer: $(6 \cdot 5) \cdot (-2) = (-2) \cdot (6 \cdot 5)$ Answer: $(2)(3 \cdot 6) = (2 \cdot 3)(6)$ Answer:	

Example	Notes
5. Multiply $\left(\dfrac{1}{4}\right)(-5)(-4)(3)$. Change the order of the terms to group simpler multiplications. Answer:	

Helpful Hints

Changing the order of factors and grouping easier multiplications together may make the arithmetic of a problem simpler to perform.

Look for numbers that multiply to powers or multiples of ten, or fractions that will cancel, in each problem before you multiply.

Multiplying a number by zero gives us a product of zero. $a \cdot 0 = 0$
Multiplying a number by one gives us the same number. $1 \cdot a = a$

Concept Check

1. What makes this multiplication easy to simplify? $-9\left(6\dfrac{3}{4}\right)(-5.023)(0)\left(-\dfrac{5}{8}\right)(20,000)$

Practice

Multiply. Change the order of the terms to group simpler multiplications.

2. $(-9 \cdot 2) \cdot (5 \cdot 6)$

3. $9(-4)\left(\dfrac{2}{3}\right)(-5)$

Which property of multiplication is shown by each equation?

4. $(-9 \cdot 3) \cdot (-12) = (-12) \cdot (-9 \cdot 3)$

5. $(-7 \cdot 2) \cdot 5 = -7(2 \cdot 5)$

Variables, Expressions, and Equations
Topic 14.1 Introduction to Expressions

Vocabulary
variable   •   term   •   algebraic expression   •   like terms   •   equation

1. A(n) _____ is a letter or symbol that is used to represent an unknown quantity.

2. A(n)_____ is any number, variable, or product of numbers and/or variables.

Step-by-Step Video Notes
Watch the Step-by-Step Video lesson and complete the examples below.

Example	Notes
1–5. Identify the terms in each expression.	
$3x$	
Determine the number of terms.	
Write the terms.	
$2x+1$	
Determine the number of terms.	
Write the terms.	
$-7xy^2 +3z-2$	
Determine the number of terms.	
Write the terms.	
60	
Determine the number of terms.	
Write the terms.	
$-6x-4z$	
Determine the number of terms.	
Write the terms.	

Example	Notes
6–10. Identify the like terms in each expression. $3x + 4y - 7x + 5z$ $2x^2 + 3x$ $-4x + 5 - 2x - 7$ $2x^2 + 3x - 1 + 6x^2 - 9x + 8$ $8a - 7b + 3b + 2b + a$	

Helpful Hints

An algebraic expression is a combination of numbers and variables, operation symbols, and grouping symbols.

Terms are the parts of an algebraic expression separated by a plus sign or a minus sign. The sign in front of the term is considered part of the term.

Like terms are terms that have the same variables raised to the same powers.

Concept Check

1. How many terms are in the algebraic expression $-6xy + 11xy^2 + 3.5$?

Practice

Identify the terms in each expression.

2. $4q + u$

3. $-7ab^2 + 8c - 11$

Identify the like terms in each expression.

4. $7x + 4y - 3x + 9y - 5$

5. $1.3x^3y + 5.6x + 3.2x^3y - 6.1x$

Variables, Expressions, and Equations
Topic 14.2 Evaluating Algebraic Expressions

Vocabulary
value   •   evaluate   •   order of operations   •   variable

1. When you _____ an expression for a given value, you are finding the value of the expression for a given value of the variables.

Step-by-Step Video Notes
Watch the Step-by-Step Video lesson and complete the examples below.

Example	Notes
1. Evaluate $3x+6$ for $x=4$. Substitute the given value for the variable. $3(4)+6$ Simplify. Remember to follow the rules for order of operations. $\boxed{}+6$ Answer:	
2. Evaluate $5x^2$ for $x=6$. Substitute the given value for the variable. $5\left(\boxed{}\right)^2$ Simplify. Remember to follow the rules for order of operations. Answer:	

Example	Notes
3. Evaluate $3y^2 - y$ for $y = -7$. Answer:	
5. The perimeter of a rectangle can be found by the expression $2(l+w)$, where l is the length and w is the width. Find the perimeter of this rectangle if $l = 5.4$ and $w = 8$. Answer:	

Helpful Hints

Remember to use the Order of Operations (PEMDAS) when you're evaluating and simplifying algebraic expressions.

Many geometric measurements are found by evaluating algebraic expressions (called formulas) at given values.

Be sure to consider the sign of the value being substituted when evaluating algebraic expressions.

Concept Check

1. Explain how you would evaluate $4x^2 - 2x + 4$ for $x = 1.5$.

Practice

Evaluate for $x = 8$.

2. $-3x + 14$

3. $x^2 - 8x - 1$

Evaluate for the given values of l and w.

4. $2(l+w)$ for $l = 4$ and $w = 6.2$

5. $2l + 2w$ for $l = 17$ and $w = 16$

Variables, Expressions, and Equations
Topic 14.3 Simplifying Expressions

Vocabulary
simplifying   •   like terms   •   algebraic expressions

1. You can simplify an expression by combining any _____.

Step-by-Step Video Notes
Watch the Step-by-Step Video lesson and complete the examples below.

Example	Notes
1–4. Combine like terms.	
$13x - 2x = \boxed{}\, x$	
$-4a + 3b + 9a = \boxed{}\, a + \boxed{}$	
$2x^3 + 9x^2 - x + 7$	
$\dfrac{4}{3}m + \dfrac{2}{3}m - m$	

5. Combine like terms.

$2a + 32b - 25c - 12b + 15a + 13c$

Rearrange the terms to group the like terms together.

Combine the like terms.

Answer:

Example	Notes
6. Combine like terms. $15x^3 + 2 - 8x^2 + x^3 - 9x + 13x^2 - 3$ Rearrange the terms to group the like terms together. Combine the like terms. Answer:	
7. Combine like terms. $16x^2y - 3xy^2 + 5xy - 2x^2y - 4xy^2$ Answer:	

Helpful Hints

If an expression has several terms, you might find it helpful to rearrange the terms before simplifying so that the like terms are together.

Remember that the sign in front of the term is part of the term. When you rearrange the terms, the sign in front of the term stays with the term.

The order in which you write the terms in the answer does not matter. However, it's customary to write the term with the highest exponent first, and number terms last.

Concept Check

1. Can you combine $2x^2y^3$ and $-5x^3y^2$? Explain.

Practice

Combine like terms.

2. $-5x + 24x$

3. $5x^3 + 3x^2 - 7x + 4$

4. $5x^3 + 3x^2 - 7x^3 + 4 - 2x^2$

5. $4 - 3y + 7y^2 + 9 - 5y^2 - 4 + 8y$

Variables, Expressions, and Equations
Topic 14.4 Simplifying Expressions with Parentheses

Vocabulary

parentheses   •   grouping symbols   •   Distributive Property

1. If an algebraic expression contains parentheses and it is not possible to simplify what is inside the parentheses first, then you can apply the _____ to remove the parentheses.

Step-by-Step Video Notes
Watch the Step-by-Step Video lesson and complete the examples below.

Example	Notes
1. Simplify $5+4(a+b)$. Use the Distributive Property to remove the parentheses. $5+4a+\boxed{}$ Combine like terms. Answer:	
2. Simplify $-(4x-3y)$. Remove the parentheses by multiplying what is inside the parentheses by -1. Answer:	
6. Simplify $14x-2\big[3x+3(5)\big]$. Remove innermost parentheses first. Use the Distributive Property to remove the brackets. Then combine like terms. Answer:	

Example	Notes
7. Simplify. $$\dfrac{7-(4-x)}{3+2(x-5)}$$ Answer:	

Helpful Hints

A negative sign in front of parentheses means the opposite of what is inside the parentheses. Remove the parentheses by multiplying what is inside the parentheses by −1.

Work "inside out." Start with the innermost grouping symbols, and remove each set of grouping symbols in turn working from the inside to the outside.

When simplifying expressions within a fraction, simplify the expressions above and below the fraction bar first. Then simplify the fraction, if possible.

Concept Check

1. Explain how you would simplify $2\left[-4(9x-7)\right]$.

Practice

Simplify.

2. $8(x+y)+4(m+n)$

4. $5x+4\left[2x+6(3x-1)\right]$

3. $-(8a+3y-2)$

5. $\dfrac{4(-3y+9)}{7(5x-6)}$

Name: _____   Date: _____

Instructor: _____   Section: _____

Variables, Expressions, and Equations
Topic 14.5 Translating Words into Symbols

Vocabulary
difference   •   quotient   •   Commutative Property

1. The order in which you write a subtraction or division expression matters, because the
   _____ does not work for subtraction or division.

2. Phrases such as "a number decreased by 6", "6 fewer than a number" and
   "the _____ between a number and 6" all indicate subtraction.

Step-by-Step Video Notes
Watch the Step-by-Step Video lesson and complete the examples below.

Example	Notes
1–3. Translate into an algebraic expression.	
The sum of 8 and a number	
Triple a number	
75% of a number	
4–7. Translate into an algebraic expression.	
Five less than twelve	
Twelve less than five	
Fifty divided by one	
One divided by fifty	

Example	Notes
11–13. Translate into an algebraic expression. Use parentheses if necessary. Seven more than double a number x A number x is tripled and then increased by 8 One-half the sum of a number x and 3	

14. Use an expression to describe the measure of each angle.

The measure of the second angle of a triangle is double the measure of the first angle, and the third angle is 15° more than the measure of the second angle.

Helpful Hints

The order of the terms is important in subtraction and division expressions.

Use parentheses when writing expressions to be sure certain operations are performed first.

Concept Check
1. What are some words or phrases in an expression that indicate addition?

Practice
Translate into an algebraic expression.
2. A number n increased by seven

4. Triple the difference of a and b

3. The quotient of 24 and some number y

5. 100 less than the product of 3 and x

Introduction to Solving Linear Equations
Topic 15.1 Translating Words into Equations

Vocabulary
variable   •   equation   •   expression

1. A(n) _____ is a letter or symbol that is used to represent an unknown quantity.

2. A(n) _____ is a mathematical statement that two expressions are equal.

Step-by-Step Video Notes
Watch the Step-by-Step Video lesson and complete the examples below.

Example	Notes
1. Translate into an equation. Let n represent the number. One-third of a number is fourteen. Look for key words to help you translate the words into algebraic symbols and expressions. One-third of a number is fourteen. $$\frac{\square}{\square} \cdot \square = \square$$ The equation is:	
2. Translate into an equation. Do not solve. Five more than six times a number is three hundred five. The equation is:	

Example	Notes
3. Translate into an equation. Do not solve. The larger of two numbers is three more than twice the smaller number. The sum of the numbers is thirty-nine. Write an expression to represent each unknown quantity in terms of the chosen variable. Let s stand for the smaller number. An expression for the larger number is _____. The equation is:	
4. Translate into an equation. Do not solve. The annual snowfall in Juneau, Alaska is 105.8 inches. This is 20.2 inches less than three times the annual snowfall in Boston, Massachusetts. The equation is :	

Helpful Hints

When writing an equation from a word problem, write an expression to represent each unknown quantity in terms of the variable. Use a given relationship in the problem or an appropriate formula to write an equation.

Concept Check
1. What words or phrases are typically associated with multiplication? Division? Addition? Subtraction?

Practice
Translate into an equation. Do not solve.

2. A number n increased by three is nine.

4. Twice the difference of a and 7 is eight.

3. One-fourth of a number is 16.

5. 100 less than the product of 3 and x is 5.

Introduction to Solving Linear Equations
Topic 15.2 Introduction to Linear Equations

Vocabulary
equation   •   solution   •   variable   •   linear equation in one variable   •   exponent

1. A(n) _____ of an equation is the number(s) that, when substituted for the variable(s), makes the equation true.

2. A(n) _____ is an equation that can be written in the form $Ax + B = C$ where $A, B,$ and C are real numbers and $A \neq 0$.

Step-by-Step Video Notes
Watch the Step-by-Step Video lesson and complete the examples below.

Example	**Notes**
1. Is 2 a solution of the equation $3x - 1 = 5$? Substitute 2 for x. $3\left(\square\right) - 1 \overset{?}{=} 5$ Simplify each side of the equation. $\square \overset{?}{=} 5$ Answer:	
2 & 3. Is -1 a solution of the equation $2x + 6 = -1$? Is -3 a solution to $7x - 2 = 5$? Answer:	

Example	Notes
4–6. Determine if each equation is a linear equation. $2x + 3 = 1$ $2x = 5$ $6x^2 - 3 = 4$	

Helpful Hints

The variable in a linear equation cannot have an exponent greater than 1.

To determine if a given value is a solution of an equation, substitute the given value into the equation. Simplify each side, and if the result is a true statement, that value is a solution.

Concept Check

1. Why is the equation $x(x+4) = 45$ not a linear equation?

Practice

Is 7 a solution of the equation?

2. $42 - 3x = 21$

Determine if each is a linear equation.

4. $5x + 4\dfrac{1}{3} = -9$

3. $31 = 4x + 5$

5. $11x - 18 = x^2$

Introduction to Solving Linear Equations
Topic 15.3 Using the Addition Property of Equality

Vocabulary
equivalent equations   •   solving the equation   •   Addition Property of Equality

1. The process of finding the solution(s) of an equation is called _____.

2. Equations that have exactly the same solutions are called _____.

Step-by-Step Video Notes
Watch the Step-by-Step Video lesson and complete the examples below.

Example	Notes
1. Solve $x + 16 = 20$ for x. Check your solution. Use the Addition Property to subtract 16 from both sides. Simplify. Check your solution. Solution:	
2. Solve $-14 = x - 3$ for x. Check your solution. Add ☐ to both sides of the equation. Simplify. Check your solution. Solution:	

Example	Notes
3. Solve $x - \dfrac{1}{2} = -\dfrac{5}{2}$ for x. Check your solution. Add $\dfrac{\square}{\square}$ to both sides of the equation. Solution:	
4. Solve $15 + 2 = 3 + x + 6$ for x. Check your solution. Solution:	

Helpful Hints

To solve an equation using the Addition Property, add or subtract the same number from both sides of the equation to get the variable x on a side of the equation by itself. If a number is being added to x, use subtraction. If a number is being subtracted, use addition.

Concept Check

1. Explain why the equations $23 - 5 = 4 + x + 3$ and $11 = x + 3$ are not equivalent.

Practice

Solve for x. Check your solution.

2. $x + 19 = 28$

4. $x - \dfrac{3}{5} = \dfrac{7}{5}$

3. $-23 = x - 6$

5. $-25 + 12 = -8 + x + 16$

Introduction to Solving Linear Equations
Topic 15.4 Using the Multiplication Property of Equality

Vocabulary
Addition Property of Equality   •   reciprocal   •   Multiplication Property of Equality

1. The _____ states that if both sides of an equation are multiplied by the same non-zero number, the solution does not change.

2. Dividing by a number is the same as multiplying by its _____.

Step-by-Step Video Notes
Watch the Step-by-Step Video lesson and complete the examples below.

Example	Notes
1. Solve $\dfrac{x}{3} = -15$ for x. Check your solution. Use the Multiplication Property to multiply each side of the equation by 3. Simplify. Check your solution. Solution:	
2. Solve $3x = 21$ for x. Check your solution. Divide both sides by $\Box$. Simplify. Check your solution. Solution:	

Example	Notes
3. Solve $20 = -4x$ for x. Check your solution. Solution:	
4. Solve $2x - 5x = -12$ for x. Check your solution. Simplify each side of the equation. Solution:	

Helpful Hints

To solve an equation using the Multiplication Property of Equality, multiply or divide both sides of the equation by the same number to get the variable x on a side of the equation by itself. If x is being multiplied by a number, use division. If x is being divided by a number, use multiplication.

Concept Check

1. By what number should you multiply each side of the equation $\frac{1}{6}x = -4$ to solve for x?

Practice

Solve for x. Check your solution.

2. $-32 = -4x$

3. $-0.4x = 2.8$

4. $\frac{x}{3} = -9$

5. $7x - 8x = 9$

Introduction to Solving Linear Equations
Topic 15.5 Using the Addition and Multiplication Properties Together

Vocabulary
Addition Property of Equality   •   variable term   •   Multiplication Property of Equality

1. To solve an equation of the form $ax+b=c$, we must use both the _____
   and the _____ together.

Step-by-Step Video Notes
Watch the Step-by-Step Video lesson and complete the examples below.

Example	Notes
1. Solve $5x+3=18$ for x to determine how many goals Jenny scored, and then check your solution. Use the Addition Property to subtract 3 from both sides. Use the Multiplication Property to divide both sides by 5. Solution:	
2. Solve $-\dfrac{1}{2}x+10=16$ for x. Use the Addition Property to subtract ☐ from both sides. Use the Multiplication Property to multiply both sides by ☐. Solution:	

Example	Notes
3. Solve $6x - 8 = -2$ for x. Solution:	
4. Solve $4 = -7 + 8x$ for x. Solution:	

Helpful Hints

To evaluate an equation of the form $ax + b = c$, the order of operations tells us to multiply before adding. When trying to solve the equation for x, we must undo this. That is, we must add (or subtract) first, and then multiply (or divide).

Concept Check

1. Which operation would you undo first to solve the equation $-2x + 8 = -14$ for x?

Practice

Solve for x.

2. $7x + 3 = 45$

4. $\frac{1}{6}x + 4 = -8$

3. $-22 = 3x - 7$

5. $4x - 13.2 = 14.8$

Solving More Linear Equations and Inequalities
Topic 16.1 Solving Equations with Variables on Both Sides

Vocabulary
solution   •   combining like terms   •   solving the equation

1. The process of finding the solution(s) of an equation is called _____.

Step-by-Step Video Notes
Watch the Step-by-Step Video lesson and complete the examples below.

Example	Notes
1. Solve $9x = 6x + 15$ for x. The goal is to get the variable alone on one side of the equation and numbers on the other side. Subtract $6x$ from both sides. Solution:	
2. Solve $9x + 4 = 7x - 2$ for x. The goal is to get the variable alone on one side of the equation and numbers on the other side. Subtract $\boxed{}\,x$ from both sides. Subtract $\boxed{}$ from both sides. Solution:	

Example	Notes
3. Solve $5x + 26 - 6 = 9x + 12x$ for x. Simplify each side. Get the variable terms on one side. Solution:	
4. Solve $-x + 8 - x = 3x + 10 - 3$ for x. Solution:	

Helpful Hints

Sometimes variable terms and number terms appear on both sides of the equation. If it is necessary, simplify one or both sides of the equation by combining like terms that are on the same side of the equation. Then get the variable terms on one side of the equation and the number terms on the other side.

Concept Check

1. Which term would you add to both sides of the equation $-2x - 8 = 14 - 5x$ so that the variable terms are on one side of the equation with a positive coefficient?

Practice

Solve for x.

2. $4x + 6 = 8x$

3. $9x - 22 = 3x - 4$

4. $7 - 3x + 2 = -9 + 4x - 10$

5. $-12x + 2 + 7x = 1 - 8x + 16$

Solving More Linear Equations and Inequalities
Topic 16.2 Solving Equations with Parentheses

Vocabulary
Distributive Property   •   parentheses   •   equation

1. In order to solve an equation with parentheses, simplify by using the _____
    to remove the parentheses.

Step-by-Step Video Notes
Watch the Step-by-Step Video lesson and complete the examples below.

Example	Notes
1. Solve $2(x+5)=-12$ for x. Simplify each side. $2\Box+\Box=-12$ Notice there is only one variable term. Get the number terms on the other side. Get the variable alone on one side. Solution:	
3. Solve $5(x+1)-3(x-3)=17$ for x. Simplify each side. $\Box x+\Box-\Box x+\Box=17$ Get the variable terms on one side. Get the number terms on the other side. Get the variable alone on one side. Solution:	

Example	Notes
5. Solve $3(0.5x - 4.2) = 0.6(x - 12)$ for x.	
Solution:	
6. Solve $2(18x - 5) + 2 = 24x - 3(12x + 8)$ for x.	
Solution:	

Helpful Hints

Recall that the Distributive Property states that for all real numbers $a, b,$ and c, $a(b + c) = ab + ac$. Sometimes an equation has multiple sets of parentheses. If this is the case, apply the Distributive Property as many times as is necessary to remove all sets of parentheses. Also, remember that parentheses can appear inside other parentheses.

Concept Check

1. Can you solve the equation $3(x - 7) = 12$ without using the Distributive Property?

Practice

Solve for x.

2. $4(x + 6) = -8$

3. $7(x - 2) - 6 = x + 4$

4. $7(-3x + 2) = -8(4x - 10)$

5. $0.3x - 2(x - 1.2) = -0.7(x - 3) - 3.7$

Solving More Linear Equations and Inequalities
Topic 16.3 Solving Equations with Fractions

Vocabulary
Distributive Property   •   least common denominator   •   equivalent equation

1. To make the process of solving equations with fractions easier, multiply both sides of the
   equation by the _____ of all the fractions contained in the equation.

Step-by-Step Video Notes
Watch the Step-by-Step Video lesson and complete the examples below.

Example	Notes
1. Solve $\dfrac{1}{4}x - \dfrac{2}{3} = \dfrac{5}{12}x$ for x. Find the LCD of the fractions, then multiply both sides of the equation by the LCD. $\square\left(\dfrac{1}{4}x - \dfrac{2}{3}\right) = \square\left(\dfrac{5}{12}x\right)$ Use the Distributive Property. Solution:	
2. Solve $\dfrac{x}{3} + 3 = \dfrac{x}{5} - \dfrac{1}{3}$ for x. Find the LCD of the fractions, then multiply both sides of the equation by the LCD. $\square\left(\dfrac{x}{3} + 3\right) = \square\left(\dfrac{x}{5} - \dfrac{1}{3}\right)$ Solution:	

Example	Notes
3. Solve $\dfrac{x+5}{7} = \dfrac{x}{4} + \dfrac{1}{2}$ for x. Find the LCD of the fractions, then multiply both sides of the equation by the LCD. Solution:	
4. Solve $0.6x - 1.3 = 4.1$ for x. Solution:	

Helpful Hints

You can also solve an equation containing decimals in a similar way to the fraction equations. You can multiply both sides of the equation by an appropriate value to eliminate the decimal numbers and work only with integer coefficients. If the decimals are tenths, multiply by 10, if the decimals are hundredths or less, then multiply by 100, etc.

Concept Check

1. By what number would you multiply each term in the equation $0.03x - .42 = 1.2$ to work with only integer coefficients?

Practice

Solve for x.

2. $\dfrac{1}{2}x - \dfrac{2}{3} = \dfrac{5}{6}$

4. $\dfrac{x+6}{12} = \dfrac{x}{6} + \dfrac{3}{4}$

3. $\dfrac{7}{8}x - \dfrac{5}{2} = \dfrac{3}{4}x$

5. $3.6 = 4(0.6x - 0.3)$

Name: _____   Date: _____
Instructor: _____   Section: _____

Solving More Linear Equations and Inequalities
Topic 16.4 Solving a Variety of Equations

Vocabulary
identity   •   infinite number of solutions   •   no solution
contradiction   •   solving an equation

1. An equation has _____ if there is no value of x that makes the equation true.

2. An equation has an _____ if the equation is always true, no matter the value of x.

Step-by-Step Video Notes
Watch the Step-by-Step Video lesson and complete the examples below.

Example	Notes
1. Solve $3(6x-4)=4(3x+9)$ for x. Remove the parentheses using the Distributive Property. Solution:	
2. Solve $2(3x+1)=5(x-2)+3$ for x. Solution:	

Example	Notes
3. Solve $\frac{1}{3}(x-2) = \frac{1}{5}(x+4) + 2$ for x. Solution:	
4. Solve $5(x+3) = 2x - 8 + 3x$ for x. Solution:	

Helpful Hints

Checking the solution is arguably the most important step of solving an equation. There are situations where possible solutions found may not actually be solutions.

It is important to follow the order of operations when solving equations.

Concept Check

1. Why is an equation such as $2x + 9 + x = 4 + 3x + 5$ called an identity?

Practice

Solve for x.

2. $\frac{1}{4}(4x - 12) = \frac{1}{5}(10x + 5)$

3. $\frac{1}{3}(x+6) = \frac{1}{6}(x-3) + \frac{2}{3}$

4. $-6 + 4(x-3) = 11x - 5 - 7x$

5. $7(x+3) - 6 = 24 - 4x - 9 + 11x$

Solving More Linear Equations and Inequalities
Topic 16.5 Solving Equations and Formulas for a Variable

Vocabulary

formula   •   Distributive Property   •   least common denominator

1. A _____ is an equation in which variables are used to describe a relationship.

Step-by-Step Video Notes
Watch the Step-by-Step Video lesson and complete the examples below.

Example	Notes
1. Solve $5x + 2 = 17$ and $ax + b = c$ for x. Identify the variable in both equations. Notice that these equations are of the same form, except that every number in the first equation is a variable in the second equation. $$5x + 2 = 17 \qquad\qquad ax + b = c$$ $$5x + 2 - \square = 17 - \square \qquad ax + b - \square = c - \square$$ $$\frac{5x}{\square} = \frac{\square}{\square} \qquad\qquad \frac{ax}{\square} = \frac{\square}{\square}$$ $$x = \square \qquad\qquad x = \frac{\square}{\square}$$ Solution:	
2. Solve $d = rt$ for t. Divide both sides of the equation by r. Solution:	

Example	Notes
4. Solve for the specified variable. $a = \dfrac{v}{t}, \; v$ Solution:	
7. Solve for the specified variable. $5x + 3y = 6,$ solve for y Solution:	

Helpful Hints

To solve a formula or an equation for a specified variable, use the same steps for solving a linear equation except treat the specified variable as the only variable in the equation and treat the other variables as if they were numbers.

Sometimes if there are two variables in an equation, you may be asked to solve for one variable *in terms of* the other. For example, "solve $2x - 3y = 6$ for y" indicates that you are to find y in terms of x.

Concept Check

1. What would you divide by to solve $A = bh$ for h?

Practice

Solve for x.

2. $2x + 8y = 12$

3. $y = mx + b$

Solve for the specified variable.

4. $A = \dfrac{1}{2}bh,$ solve for h.

5. $8x - 3y = 12,$ solve for y.

Solving More Linear Equations and Inequalities
Topic 16.6 Solving and Graphing Linear Inequalities

Vocabulary
inequality   •   linear inequality in one variable   •   solution of an inequality
graph of an inequality   •   solve an inequality   •   non-solutions

1. The _____ is a picture that represents all of the solutions of the inequality.

2. A(n) _____ is a statement that shows the relationship between any two real numbers that are not equal.

Step-by-Step Video Notes
Watch the Step-by-Step Video lesson and complete the examples below.

Example	Notes
1 & 2. Graph each inequality on a number line.	

$x > 3$

Use a(n) _____ circle at the boundary point $x = 3$, because 3 _____ a solution.

Shade all numbers to the _____ side of the boundary point.

$x \leq -1$

Use a(n) _____ circle at the boundary point $x = -1$, because -1 _____ a solution.

Shade all numbers to the _____ side of the boundary point.

Example	Notes
3. Solve and graph the inequality $5x+2<12$. Subtract 2 from both sides of the inequality. Divide both sides of the inequality by 5. Graph the inequality. ← + + + + + + + + + + + → x −5 −4 −3 −2 −1 0 1 2 3 4 5	
4. Solve and graph the inequality $5-4x \geq -7$. ← + + + + + + + + + + + → x −5 −4 −3 −2 −1 0 1 2 3 4 5	

Helpful Hints
It is important to decide if you need an open circle or a closed circle. Remember, for the endpoint, use an open circle for $<$ or $>$ and a closed circle for $\leq$ or $\geq$.

Use the same procedure to solve an inequality that is used to solve an equation, *except* the direction of an inequality must be *reversed* if you *multiply* or *divide* both sides of the inequality *by a negative* number.

Concept Check
1. Would you use an open or closed circle for the boundary point to graph $2x+3 \leq 7$?

Practice
Graph the inequality on a number line.
2. $x<2$

Solve and graph the inequality.
4. $8+4x>6$

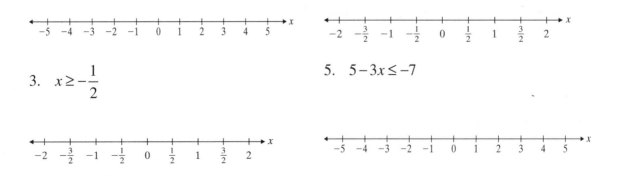

3. $x \geq -\dfrac{1}{2}$

5. $5-3x \leq -7$

Name: Sidney

Instructor: Schnabel

Date: 1/13/16

Section: _____

Introduction to Graphing Linear Equations
Topic 17.1 The Rectangular Coordinate System

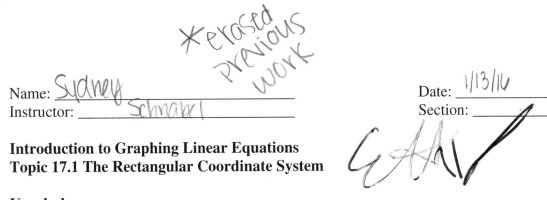

Vocabulary

rectangular coordinate system • *x*-coordinate • *y*-coordinate • ordered pair

1. A ___rectangular coordinate system___ is made up of a horizontal number line and a vertical number line that intersect to form a right angle. The point where these number lines meet is the origin.

Step-by-Step Video Notes
Watch the Step-by-Step Video lesson and complete the examples below.

Example	Notes
1. Plot the points $(4, 2), (3, -2), (-3, 3)$, and $(-1, -4)$. Label them A, B, C, and D, respectively. 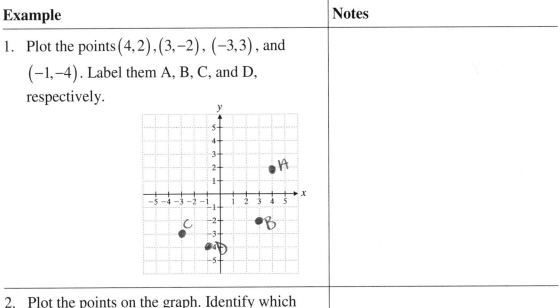	
2. Plot the points on the graph. Identify which quadrant each point lies in. Label them A, B, C, and D, respectively. a. $(4, -5)$ 4 b. $(-5, 4)$ 2 c. $(3, 0)$ none d. $(2, 2)$ 1 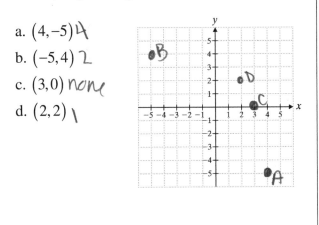	

Copyright © 2012 Pearson Education, Inc.

Example	Notes

3. Find the coordinates of the indicated points. Write each point as an ordered pair. Identify which quadrant each point lies in.

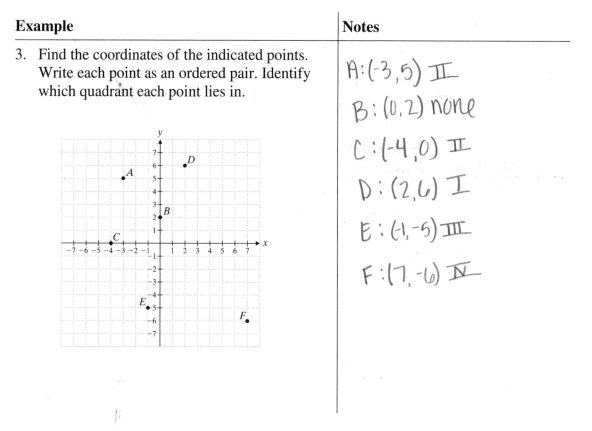

Notes (handwritten):

A: (-3, 5) II

B: (0, 2) none

C: (-4, 0) II

D: (2, 6) I

E: (-1, -5) III

F: (7, -6) IV

Helpful Hints

The origin is represented by the ordered pair $(0,0)$, and has both x- and y-coordinates of 0.

The quadrants are numbered I, II, III, and IV, starting at the top right and going counter-clockwise. Points that lie on an axis are not considered to be in any quadrant.

Concept Check

1. Give an example of a point in Quadrant IV. Give an example of a point not in a quadrant.

(5, -5) IV   not in quad (0, 0)

Practice

Plot the given points on the graph. Identify which quadrant each point lies in.

2. A$(-3, 4)$ 3. B$(4, -5)$ 4. C$(0, -2)$ 5. D$(4, 1)$

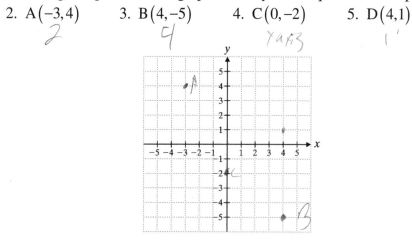

Name: Sydney

Instructor: Schnabel

Date: 1/13/16

Section: _____

Introduction to Graphing Linear Equations
Topic 17.2 Graphing Linear Equations by Plotting Points

Vocabulary
linear equations in two variables   •   solution to an equation   •   ordered pair

linear equations in two variables

1. A __solution to an equation__ is an equation that can be written in the form $Ax + By = C$
   where $A, B,$ and C are real numbers, but A and B are not both zero.

Step-by-Step Video Notes
Watch the Step-by-Step Video lesson and complete the examples below.

Sydney's Work

Example	Notes
1. Determine whether $(3,5)$ is a solution to the equation $3x + 2y = 19$. *yes* / **Yes**	*Previous* (3,5)? $3x + 2y = 19$ $3(3) + 2(5) = 19$ $9 + 10 = 19$ (yes)
	Sydney's Work $3(3) + 2(5) = 19$ $9 + 10 = 19$ ✓ Yes!

Substitute the x- and y-coordinates into the
linear equation. Check for a true statement.

$$3\left(\boxed{3}\right) + 2\left(\boxed{5}\right) \overset{?}{=} 19$$

$$\boxed{19} \overset{?}{=} 19$$

4.4
$3x + 2y = 19$
$3(4) + 2(4) = 19$
$12 + 8 = 19$
(NO)

Answer: *yes* **Yes**

---

4. Find three solutions to $2x + y = 13$.

$(2,9)$ $(1,11)$ $(3,7)$

Substitute a value for one of the variables.

$(1,11)$ $(4,5)$ $(3,7)$

Solve the equation for the other variable.

Repeat for the other two solutions.

Write the ordered pairs.

Answer: $(2,9)$ $(1,11)$ $(3,7)$

$(1,11)$ $(4,5)$ $(3,7)$

Notes column:
$2x + y = 13$
$2(2) + a = 13$
(yes)
$2 \cdot 11 = 13$
(yes)
$6 + 7 = 13$
(yes)

Sydney's Work:
$2(1) + y = 13$
$2 + y = 13$
     -2
$y = 11$
$\boxed{x=1}$ $\boxed{y=11}$

$2(3) + y = 13$
$6 + y = 13$
     -6
$y = 7$
$\boxed{x=3}$ $\boxed{y=7}$

$2(4) + y = 13$
$8 + y = 13$
     -8
$y = 5$
$\boxed{x=4}$ $\boxed{y=5}$

Example	Notes

Example

6. Find three solutions to $x + y = -4$.

Make a table of values to keep your ordered pairs organized.

$(1, -5)$ $(-2, 2)$ $(-4, 0)$

Answer: *(handwritten)* $(2,2)$ $(3,1)$ $(-6,2)$

X	Y
1	-5
-2	-2
-4	0

Notes *(handwritten)*

$x + y = -4$
$-6 + 2 = -4$
$2 + 2 = -4$

$1 + y = -4$
$ -1$
$y = -5$

$-2 + y = -4$
$ +2$
$y = -2$

$-4 + y = -4$ $y = 0$
$ +4$

7. Graph the equation $y = 2x + 1$.

Make a table of values to find three ordered pair solutions. Plot the ordered pairs on a graph. Draw a line through the points.

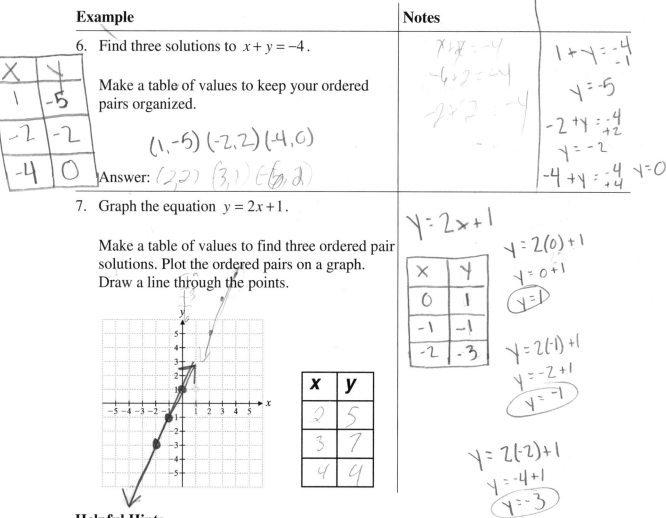

X	Y
0	1
-1	-1
-2	-3

x	**y**
2	5
3	7
4	9

(handwritten Notes)
$y = 2x + 1$

$y = 2(0) + 1$
$y = 0 + 1$
$y = 1$

$y = 2(-1) + 1$
$y = -2 + 1$
$y = -1$

$y = 2(-2) + 1$
$y = -4 + 1$
$y = -3$

Helpful Hints

To graph a linear equation in two variables, find at least 3 ordered pairs which are solutions of the equation. Plot the ordered pairs, and then draw a line through the points.

Concept Check

1. How many ordered pairs are solutions to the equation $37.5x - 19.2y = 4.8$?

(handwritten) 2

Practice

Is the given point a solution of $2x + 8y = 20$?

2. $(2, 2)$

(handwritten) yes

3. $(-6, 4)$

(handwritten) no

Find three solutions to the given equation.

4. $2x - y = 7$

(handwritten) $(7, 7)$

5. $y = -4x + 1$

(handwritten) $(2, -7)$
$(3, -11)$
$(4, -15)$

Introduction to Graphing Linear Equations
Topic 17.3 Graphing Linear Equations Using Intercepts

Vocabulary

~~intercept~~   •   ~~x-intercept~~   • y-intercept   •   origin

1. The point at which a line crosses an axis is called a(n) ___intercept___. **intercept**

2. The ___x-intercept___ is an ordered pair with the coordinates $(a, 0)$, where a is a real number. **x-intercept**

Step-by-Step Video Notes
Watch the Step-by-Step Video lesson and complete the examples below.

Example	Notes ✳ Sydney's work
1. Find the x-intercept and y-intercept of $3x - 6y = 12$. Find the x-intercept by letting $y = 0$ and solving for x. Find the y-intercept by letting $x = 0$ and solving for y. Answer: $(4, -2)$ $(4, -2)$	$3x - 6(0) = 12$ $3(0) - 6y = 12$ $3x = 12$ $-6y = 12$ $\boxed{x = 4}$ $\boxed{y = -2}$ $(4, 0)$ $(0, -2)$
2. Find the x-intercept and y-intercept of $y = \dfrac{4}{5}x$. Find the x-intercept by letting $y = 0$ and solving for x. $x = (0, 0)$ Find the y-intercept by letting $x = 0$ and solving for y. $y = (0, 0)$ Answer: $(0, 0)$ $(0, 0)$	$0 = \dfrac{4}{5}x \cdot \dfrac{5}{4}$ $0 = 1x$ $y = \dfrac{4}{5}(0)$ $\boxed{x = 0}$ $\boxed{y = 0}$

Example	Notes

3. Graph $2x - y = 4$ using the intercepts.

Make a table of values to find the *x*-intercept, *y*-intercept, and another value.

Plot the points on the graph and draw a line through the points.

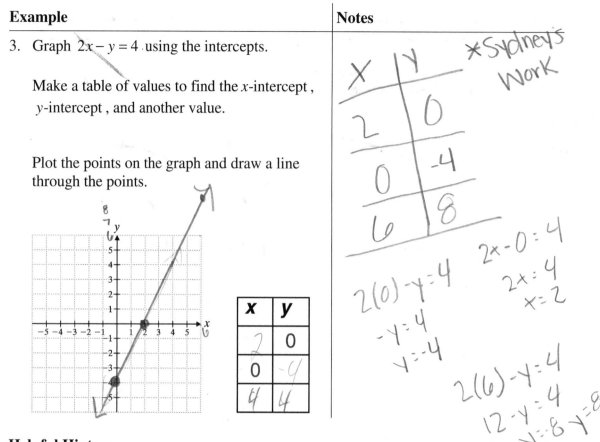

x	**y**
2	0
0	-4
4	4

Sydney's Work

X	Y
2	0
0	-4
6	8

$2(0) - y = 4$
$-y = 4$
$y = -4$

$2x - 0 = 4$
$2x = 4$
$x = 2$

$2(6) - y = 4$
$12 - y = 4$
$-y = -8$ $y = 8$

Helpful Hints

Graphing an equation using the intercepts is exactly the same process as graphing lines by plotting points. In this method, you use two specific points, (the *x*-intercept and the *y*-intercept) then plot one more ordered pair and draw a line through the points.

Concept Check

1. Give an example of a linear equation in two variables where both intercepts are the origin, $(0,0)$. $2 - y = 3x \cdot 2$

Practice

Find the *x*- and *y*-intercepts of each equation.

2. $5x - 4y = 20$

$(-5, 4)$

3. $y = \dfrac{1}{2}x + 3$

$(-6, 3)$

4. $2 - y = 3x + 2$
$2 - y = 2$
$(0,0)$

5. $y = -4x + 1$
$(-1, 1)$

Name: _Sydney_ Date: _4/14/16_
Instructor: _Schrabel_ Section: _____

Introduction to Graphing Linear Equations
Topic 17.4 Graphing Linear Equations of the Form $x = a$, $y = b$, $y = mx$

Vocabulary
origin   •   horizontal line   •   vertical line

1. If an equation is of the form $y = b$, where b is some real number, then the graph of the equation is a ___horizontal line___ _horizontal line_
2. If an equation is of the form $x = a$, where a is some real number, then the graph of the equation is a ___vertical___ _line_. _vertical line_

Step-by-Step Video Notes
Watch the Step-by-Step Video lesson and complete the examples below.

Example	Notes

★ Sydneys Work

1. Graph $6x - 2y = 0$ using three points.
   $3, 9$

   Make a table of values. Then, plot the points on a graph and draw a line through the points.

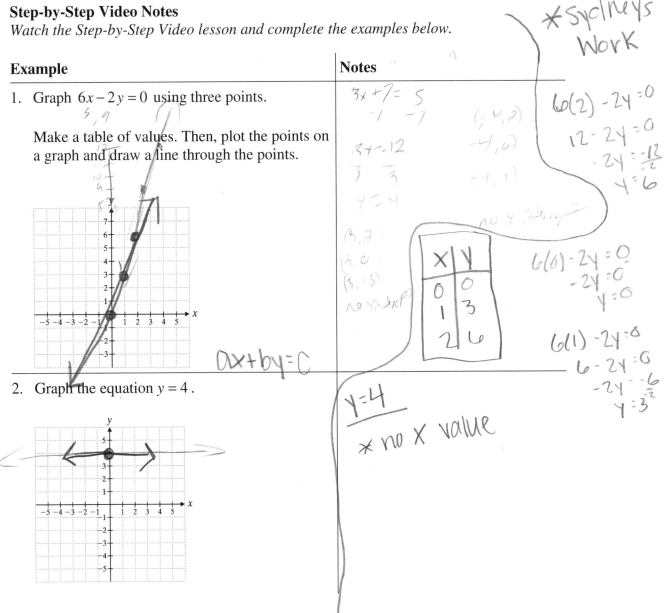

$3x + 7 = 5$
$-7 \quad -7$ $(; 4, 2)$

$3x = 12$ $(-4, 0)$
$\frac{3x}{3} = \frac{12}{3}$
$x = 4$ $(4, 1)$

$(3, 2)$
$(3, 0)$ no y intercept
$(3, -3)$
no y intercept

$ax + by = c$

X	Y
0	0
1	3
2	6

$6(2) - 2y = 0$
$12 - 2y = 0$
$-2y = \frac{-12}{-2}$
$y = 6$

$6(0) - 2y = 0$
$-2y = 0$
$y = 0$

$6(1) - 2y = 0$
$6 - 2y = 0$
$-2y = \frac{-6}{-2}$
$y = 3$

2. Graph the equation $y = 4$.

$y = 4$
$★$ no x value

223

Example	Notes

5. Graph $3x + 7 = -5$

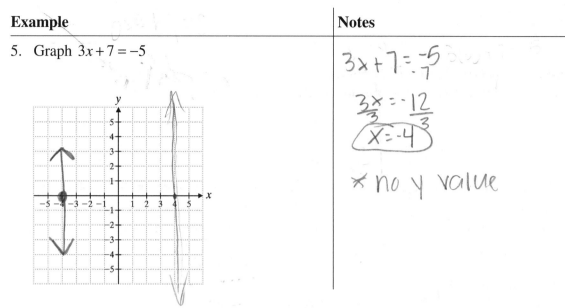

$3x + 7 = -5$
$\quad\quad -7\quad -7$

$\dfrac{3x}{3} = \dfrac{-12}{3}$

$\boxed{x = -4}$

✗ no y value

Helpful Hints

The graph of the vertical line has no y-intercept, unless its equation is $x = 0$.

Similarly, the graph of a horizontal line has no x-intercept, unless its equation is $y = 0$.

Any equation of the form $y = mx$ is neither vertical nor horizontal, and its x- and y-intercepts are the origin, $(0, 0)$.

Concept Check

1. The graph of what equation would include all of the points on the $x-$ axis ? What are its intercepts?
$x = A \quad (A, 0)$

Practice

Graph each equation.

2. $2x - 4y = 0$ **3.** $y - 1 = 2$ **4.** $x = -2$

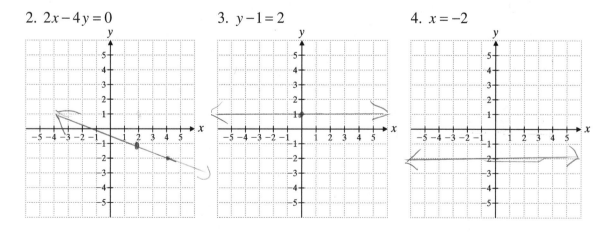

Name: Sydney Schnabel

Date: 1/14/16

Instructor:

Section:

Slope and the Equation of a Line
Topic 18.1 The Slope of a Line

Vocabulary

rise • run • slope • ordered pair

1. The _____run_____ of a line is the change in horizontal position, or the difference between the x-coordinates of two points on the line.

2. The _____slope_____ of a line is the rate of change between any two ordered pair solutions to a linear equation. Slope

Step-by-Step Video Notes

Watch the Step-by-Step Video lesson and complete the examples below.

$M = \dfrac{rise}{run}$ slope

Example	Notes
1. Find the slope. $\dfrac{rise\ 3}{run\ 2}$ $\dfrac{rise = 3}{run = 2}$ Answer: $\dfrac{3}{2}$ | $\dfrac{3}{2}$	$\dfrac{rise}{run}$ rise is the distance up and down run the distance left to right
3. Find the slope. $\quad (-4,4)$ $\quad (0,2)$ $rise = (-4 \cdot 0) = -4$ $run = (4 \cdot 2) = 2$ $m = \dfrac{-4}{2} = -\dfrac{1}{2}$ Answer: $\dfrac{1}{2}$ $\quad m = -\dfrac{1}{2}$	negative slope means goes down from left to right. Slope $= \dfrac{\text{change in } y}{\text{change in } x}$ $\dfrac{y_2 - y_1}{x_2 - x_1}$ y_1/x_2

225

Copyright © 2012 Pearson Education, Inc.

Example	Notes

5. Find the slope of the line which contains the points $(1,3)$ and $(5,11)$. Use the slope formula

$$m = \frac{y_2 - y_1}{x_2 - x_1}.$$

$$m = \frac{\boxed{~} - \boxed{~}}{\boxed{~} - \boxed{~}}$$

Answer:

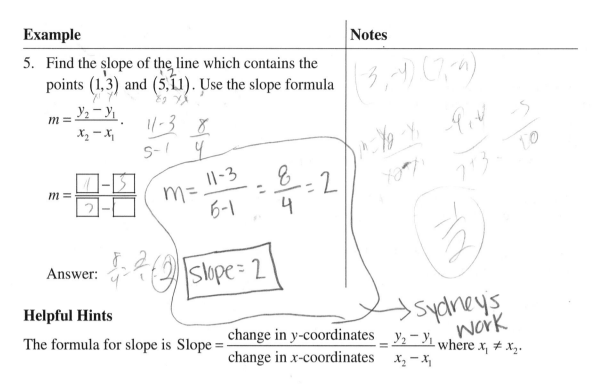

Notes (handwritten): $(-3,-4)(7,-4)$ $m = \frac{y_2 - y_1}{x_2 - x_1}$ → Sydneys work

Example work (handwritten): $m = \frac{11-3}{5-1} = \frac{8}{4} = 2$, Slope = 2

Helpful Hints

The formula for slope is $\text{Slope} = \dfrac{\text{change in } y\text{-coordinates}}{\text{change in } x\text{-coordinates}} = \dfrac{y_2 - y_1}{x_2 - x_1}$ where $x_1 \neq x_2$.

The variable m is typically used to represent slope.

A line with a positive slope goes up from left to right. A line with a negative slope goes down from left to right. The slope of a horizontal line is zero. The slope of a vertical line is undefined; it can also be said to have no slope. No slope is not the same as zero slope.

Concept Check
1. A line's run is positive and its rise is negative. Is the slope of this line positive or negative? negative

Practice
Find the slopes of the lines containing the given points.

2. $(2,8)$ and $(0,0)$

$$\frac{0-8}{0-2} = \frac{-8}{-2} \quad \boxed{4}$$

3. $(3,1)$ and $(-5,5)$

$$\frac{5-1}{-5-3} = \frac{4}{-8} = -\frac{1}{2}$$

Find the slopes of the lines.

4. $y = 4$

$\boxed{0}$

5. $x = -2.5$

undefined

Name: Sydney

Instructor: Schnabel

Date: 1/15/16

Section:

Slope and the Equation of a Line
Topic 18.2 Slope-Intercept Form

Vocabulary

slope-intercept form of a linear equation • y-intercept • Standard Form

1. The ___Slope-intercept form of a___ that has a slope m and a y-intercept $(0,b)$
   is given by the formula $y = mx + b$. linear equation

Step-by-Step Video Notes

Watch the Step-by-Step Video lesson and complete the examples below.

Example	Notes
1. Find the slope and y-intercept of $y = \dfrac{2}{3}x + 5.$ $(0,5)$ Find m. $\boxed{\dfrac{2}{3}}$ Find b. $\boxed{5}$ Answer:	$y = mx + b$ $\underset{slope}{\uparrow} \qquad \underset{y\text{-intercept}}{\uparrow}$
3. Find the slope and y-intercept of $4x - 3y = 12.$ Rewrite the equation in slope-intercept form. $m = \dfrac{4}{3}$ $y\text{-intercept} = -4$ Answer: $y = \dfrac{4}{3}x - 4$	$4x - 3y = 12$ $\dfrac{-3y}{-3} = \dfrac{-4x + 12}{-3}$ $y = \dfrac{4}{3}x - 4 \quad (0,-4)$ $\qquad \underset{m}{\uparrow} \qquad \underset{b}{\uparrow}$
6. Find the equation of a line with a slope of 2 and a y-intercept of $\left(0, \dfrac{4}{3}\right)$. Substitute the values for slope and y-intercept into the slope-intercept form. Answer: $y = 2x + \dfrac{4}{3}$	$y = 2x + \dfrac{4}{3}$

Example	Notes

8. Write the equation of the line shown in the graph.

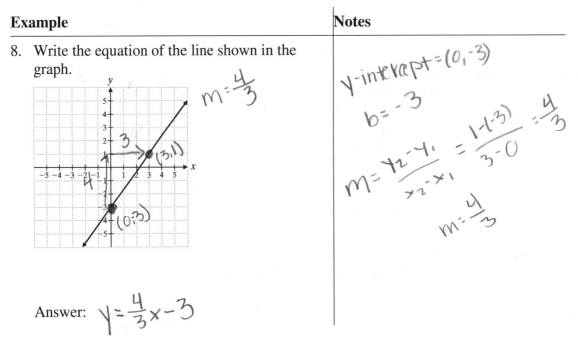

$m = \frac{4}{3}$

y-intercept $= (0, -3)$

$b = -3$

$m = \frac{y_2 - y_1}{x_2 - x_1} = \frac{1-(-3)}{3-0} = \frac{4}{3}$

$m = \frac{4}{3}$

Answer: $y = \frac{4}{3}x - 3$

Helpful Hints

You can write an equation in slope-intercept form by solving it for y.

Horizontal lines of the form $y = b$ are simplified from $y = 0x + b$, the slope-intercept form of a line with zero slope. Vertical lines of the form $x = a$ cannot be written in slope-intercept form since they have an undefined slope.

Concept Check

1. How do you know the graph of the equation $y = -3x + 5$ passes through the point $(0, 5)$?

   the y intercept is five

Practice

Find the slope and y-intercept of the given equation.

2. $x - 2y = 6$

   $-2y = -x + 6$

   $y = \frac{x}{2} - 6$

   $m = \frac{1}{2}$

   $b = -6$

3. $y = 3x - 4$

   $m = 3$

   $b = -4$

Write the equation of a line with the given slope and y-intercept.

4. $m = 0.5$, $b = -2$

   $y = \frac{1}{2}x - 2$

5. $m = -\frac{3}{4}$, $b = 6$

   $y = -\frac{3}{4}x + 6$

Name: Sydney

Instructor: Schnabel

Date: 1/15/16

Section: ✓

Slope and the Equation of a Line
Topic 18.3 Graphing a Line Using Slope and *y*-Intercept

Vocabulary

slope • *y*-intercept • rise • run

1. The __Slope__ of a line can be described as "the rise over the run."

Step-by-Step Video Notes
Watch the Step-by-Step Video lesson and complete the examples below.

Example	Notes
1. Find the *y*-intercept of the graph. 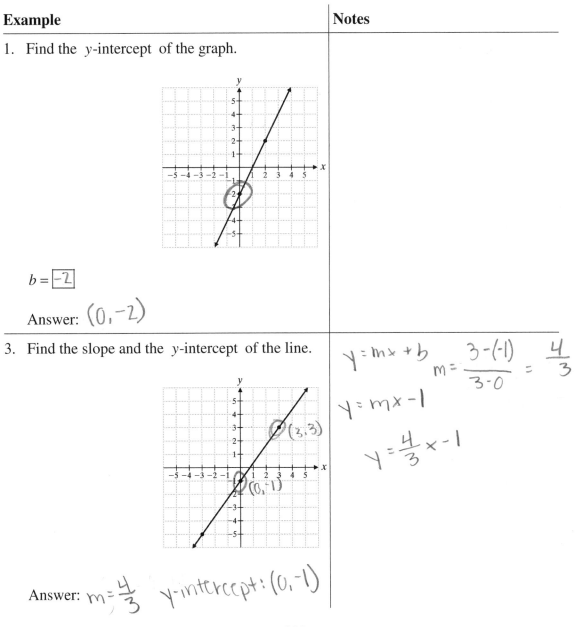 $b = \boxed{-2}$ Answer: $(0, -2)$	
3. Find the slope and the *y*-intercept of the line.	$y = mx + b$ $m = \dfrac{3-(-1)}{3-0} = \dfrac{4}{3}$ $y = mx - 1$ $y = \dfrac{4}{3}x - 1$

(3,3)

(0,-1)

Answer: $m = \dfrac{4}{3}$ *y*-intercept: $(0, -1)$

229

Example	**Notes**

4. Graph $y = \dfrac{1}{2}x - 3$.

The y-intercept is $\boxed{-3}$.

The slope is $\boxed{\dfrac{1}{2}}$.

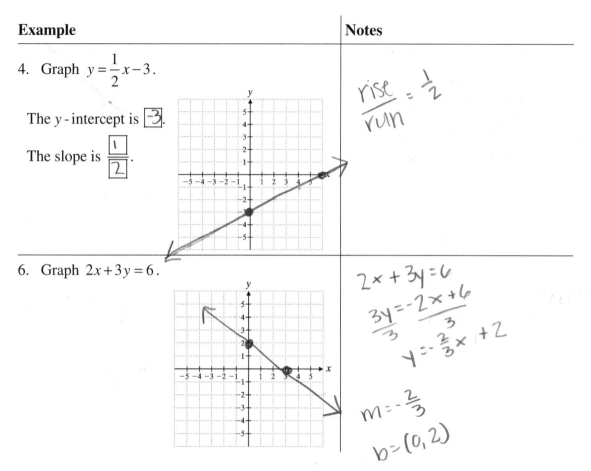

$\dfrac{\text{rise}}{\text{run}} = \dfrac{1}{2}$

6. Graph $2x + 3y = 6$.

$2x + 3y = 6$

$\dfrac{3y}{3} = \dfrac{-2x + 6}{3}$

$y = -\dfrac{2}{3}x + 2$

$m = -\dfrac{2}{3}$

$b = (0, 2)$

Helpful Hints
If the slope is an integer, its denominator is 1. If x has no coefficient in slope-intercept form, the slope is 1. If there is no constant b in slope-intercept form, the y-intercept is 0.

Concept Check

1. A line has a y-intercept of -2. Explain how to plot another point if the slope is $-\dfrac{2}{3}$.

Practice
Graph each equation.

2. $y = x + 1$

3. $y = \dfrac{1}{3}x - 1$

4. $y = -2x + 4$

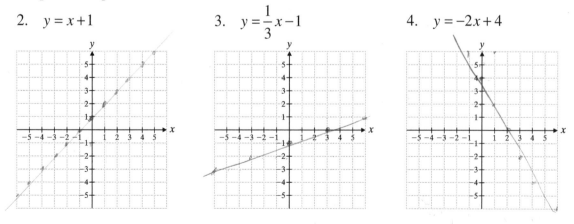

Name: Sydney   Date: 1/17/16

Instructor: Schnabel   Section: _____

✓

Slope and the Equation of a Line
Topic 18.4 Writing Equations of Lines Using a Point and Slope

Vocabulary

Slope-intercept form of a linear equation   •   point-slope form   •   standard form

1. The point slope form of an equation of a line whose slope is m and passes
through the point (x_1, y_1) is given by $y - y_1 = m(x - x_1)$.

Step-by-Step Video Notes

Watch the Step-by-Step Video lesson and complete the examples below.

Example	Notes
1. Use the point-slope form to write the equation of the line with a slope of 2 and passes through $(2,1)$. Write the point-slope equation. $y - y_1 = m(x - x_1)$ Substitute the given values of x_1, y_1, and m into the equation. $y - 1 = 2(x - 2)$ Simplify. Answer: $y - 1 = 2(x - 2)$	$y - y_1 = m(x - x_1)$ $m = 2$ $(2,1)$ $x_1\ y_1$

2. Write the equation of the line with a slope of -3 and passes through $(3,9)$. Write your answer in slope-intercept form.
 	
 Write the point-slope equation.
 $y - 9 = -3(x - 3)$
 	
 Solve the equation for y.
 	
 Answer: $y = -3x + 18$
	$m = -3$ $(3,9)$
	   $x_1\ y_1$
	
	 $y - 9 = -3x + 9$
	   $+9\quad\quad +9$
	 $y = -3x + 18$

Example	Notes
3. Write the equation of the line with a slope of $\frac{1}{7}$ and passes through $(14, -5)$. Write your answer in slope-intercept form. Answer: $y = \frac{1}{7}x - 7$	$m = \frac{1}{7}$ $\quad y + 5 = \frac{1}{7}(x - 14)$ $(14, -5)$ $\quad x_1 \quad y_1$ $y + 5 = \frac{1}{7}x - 2$ $\quad -5 \qquad\qquad -5$ $y = \frac{1}{7}x - 7$
4. Use the point-slope form to write the equation of the line with a slope of 0 and passes through $(-16, -9)$. $y + 9 = 0$ $\quad -9 \quad -9$ Answer: $y = -9$	$m = 0 \quad (-16, -9)$ $\quad\quad\quad x_1 \quad y_1$ $y - (-9) = 0(x - (-16))$ $y + 9 = 0(x + 16)$

Helpful Hints

Point-slope form tells the slope of the line and the coordinates of a point on the line (not necessarily the y-intercept).

Linear equations are usually not left in point-slope form. Most of the time, they will be expressed in slope-intercept form.

Concept Check

1. What is the slope of the graph of the equation $y - 32.7 = 1.5(x - 41.3)$?

1.5

Practice

Write the equation of the line with the given slope and passes through the given point. Leave your answer in point-slope form.

2. slope of -4 and passes through $(-6, 5)$

$y - 5 = -4(x + 6)$

3. slope of $\frac{3}{7}$ and passes through $\left(2, -\frac{8}{9}\right)$

$y + \frac{8}{9} = \frac{3}{7}(x - 2)$

Write the equation of the line with the given slope and passes through the given point. Write your answer in slope-intercept form.

4. slope of 3 and passes through $(5, 7)$

$y - 7 = 3(x - 5)$
$y = 3x - 8$

5. slope of $\frac{1}{2}$ and passes through $(-2, -1)$

$y + 1 = \frac{1}{2}(x + 2)$
$y = \frac{1}{2}x$

Slope and the Equation of a Line
Topic 18.5 Writing Equations of Lines Using Two Points

Vocabulary

slope-intercept form   •   point-slope form   •   intercept

1. To write the equation of a line when given two points, use the points to find the slope, then pick one of the points and write the equation in _point-slope form_

Step-by-Step Video Notes

Watch the Step-by-Step Video lesson and complete the examples below.

Example	Notes
1. Write the equation in slope-intercept form of the line which passes through $(2,5)$ and $(6,3)$. $x_1, y_1 \quad x_2, y_2$ Find the slope m. $m = -\frac{1}{2}$ Substitute the values of x_1, y_1, and m into the point-slope form. $y - 5 = -\frac{1}{2}(x - 2)$ Solve for y to write the answer in slope-intercept form. Answer: $y = -\frac{1}{2}x + 6$	$m = \dfrac{y_2 - y_1}{x_2 - x_1} = \dfrac{3-5}{6-2} = \dfrac{-2}{4} = -\dfrac{1}{2}$ $m = -\frac{1}{2}$ $y - 5 = -\frac{1}{2}(x-2)$ $y - 5 = -\frac{1}{2}x + 1$ $ + 5 \qquad\qquad + 6$ $y = -\frac{1}{2}x + 6$
3. Write the equation of the line which passes through $(-5,5)$ and $(0,-6)$. Write your answer in slope-intercept form. $x_1 \ y_1 \quad x_2 \ y_2$ Answer: $y = -\frac{11}{5}x - 6$	$m = \dfrac{-6-5}{0+5} = \dfrac{-11}{5}$ $m = -\frac{11}{5}$ $y - 5 = -\frac{11}{5}(x+5)$ $y - 5 = -\frac{11}{5}x - 11$ $ + 5 \qquad\qquad + 5$ $y = -\frac{11}{5}x - 6$

233

Example	Notes
4. Write the equation of the line which passes through $(-8,4)$ and $(-5,0)$. Write your answer in slope-intercept form. Answer: $y=-\frac{4}{3}x-\frac{20}{3}$	$m=\frac{0-4}{-5+8}=-\frac{4}{3}$ $\frac{4}{3}\cdot\frac{8}{1}=\frac{32}{3}$ $y-4=-\frac{4}{3}(x+8)$ $y-4=-\frac{4}{3}x-\frac{32}{3}$ $-\frac{32}{3}+(\frac{4}{1})3$ $\quad+4 \qquad +4$ $-\frac{32}{3}+\frac{12}{3}$ $y=-\frac{4}{3}x-\frac{20}{3}$
5. Write the equation of the line on the graph. Write your answer in slope-intercept form. Answer: $y=\frac{3}{4}x+\frac{5}{2}$	$m=\frac{4-1}{2+2}=\frac{3}{4}$ $\frac{3}{4}\cdot\frac{2}{1}=\frac{6}{4}=\frac{3}{2}$ $y-1=\frac{3}{4}(x+2)$ $\frac{3}{2}+(\frac{4}{1})2$ $y-1=\frac{3}{4}x+\frac{3}{2}$ $\frac{3}{2}+\frac{2}{1}=\frac{5}{2}$ $\quad+1$ $y=\frac{3}{4}x+\frac{5}{2}$

Helpful Hints

Sometimes when given two points to find a line, one of the points is the y-intercept. If this is the case, you can find the slope and then use that point to write the equation in slope-intercept form directly rather than using point-slope form first.

Concept Check

1. Find the slope-intercept form of the line through the points $(-6,5)$ and $(0,3)$ without using point-slope form. $\frac{3-5}{0+6}$ $\frac{-2}{6}$ $-\frac{1}{3}$

Practice

Write the equation of the line that passes through the given points.

2. $(-2,-5)$ and $(8,5)$

4. $(-11,-4)$ and $(9,8)$

3. $(-8,13)$ and $(-6,5)$

5. $(-2,-2)$ and $(10,4)$

Name: _Sydney_   Date: _2/8/16_

Instructor: _Schnabel_   Section: _____

Introduction to Functions
Topic 19.1 Relations and Functions

_LS_

Vocabulary

relation • domain • range • function • ordered pair

1. The second coordinates, or y values, in all of the ordered pairs of a relation make up the _range_ of the relation.

2. A _ordered pairs_ is a relation for which every x value in the domain has one and only one y value.

Step-by-Step Video Notes
Watch the Step-by-Step Video lesson and complete the examples below.

Example	Notes
1. State the domain and range of the relation. $\{(5,7),(9,11),(10,7),(12,14)\}$ State the domain. $5, 9, 10, 12$ State the range. $7, 11, 14$ Answer:	$D: \{5, 9, 10, 12\}$ $R: \{7, 11, 14\}$

2 & 3. Determine whether each relation is a function.
 $\{(3,9),(4,16),(5,9),(6,36)\}$
     yes
 	
 $\{(7,8),(9,10),(12,13),(7,14)\}$.
     no
	2. yes
	
	 3. no

Example	**Notes**

4–6. Determine whether each relation is a function.

$y = 3x - 5$

x	-2	1	4	7
y	1	2	3	4

$y = |x|$ yes

x		2	3	4
y	1	2	3	4

$y^2 = x$

x	3	4	2	6
y	9	16	4	36

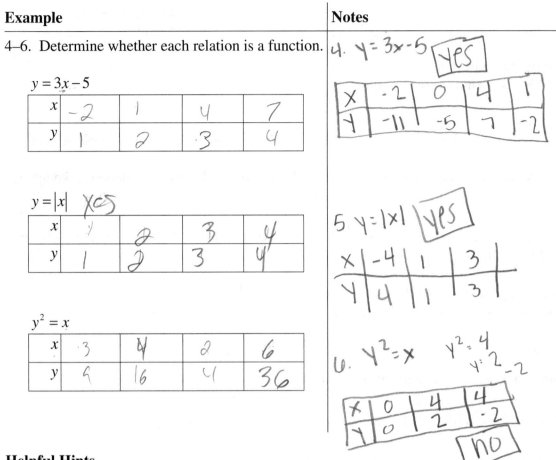

4. $y = 3x - 5$ yes

x	-2	0	4	1
y	-11	-5	7	-2

5. $y = |x|$ yes

x	-4	1	3
y	4	1	3

6. $y^2 = x$ $y^2 = 4$ $y = 2$, -2

x	0	4	4
y	0	2	-2

no

Helpful Hints

A relation is any set of ordered pairs. Some relations cannot be expressed by an equation.

The first coordinates, or *x* values, in all of the ordered pairs of a relation make up the domain of the relation.

Concept Check

1. If you switch the order of the ordered pairs in the relation $\{(-2,4),(-1,1),(1,1),(2,4)\}$, will it still be a function? Explain.

Practice

State the domain and range of the relation.

2. $(8,5)\{(3,9),(9,3),(4,6),(6,4)\}$

Determine whether each relation is a function

4. $x = |y|$

3. $y = x^2 - 1$

5. $y = x^2 + 4$

Introduction to Functions
Topic 19.2 The Vertical Line Test

Vocabulary

vertical line test   •   *x*-axis   •   ordered pairs

1. The __vertical line test__ states that if a vertical line can pass along the *x*-axis and cross the graph in at most one place, then the graph represents a function.

Step-by-Step Video Notes
Watch the Step-by-Step Video lesson and complete the examples below.

Example	Notes
1. Determine whether the following is the graph of a function. *yes* Answer:	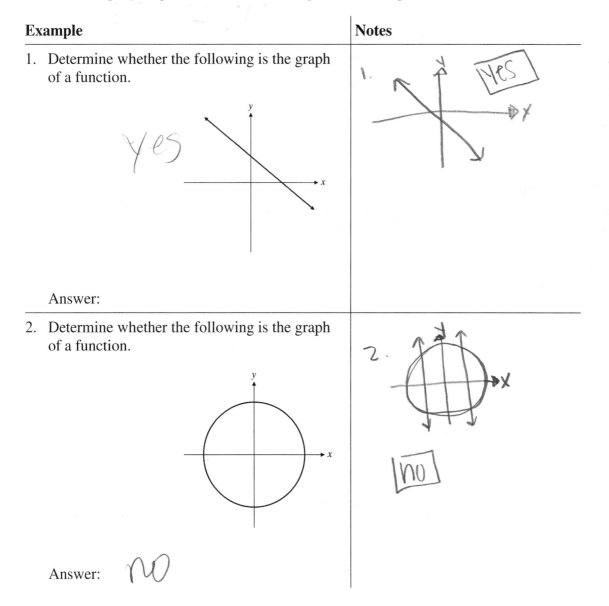
2. Determine whether the following is the graph of a function. Answer: *no*	

Example	**Notes**

3. Determine whether the following is the graph of a function.

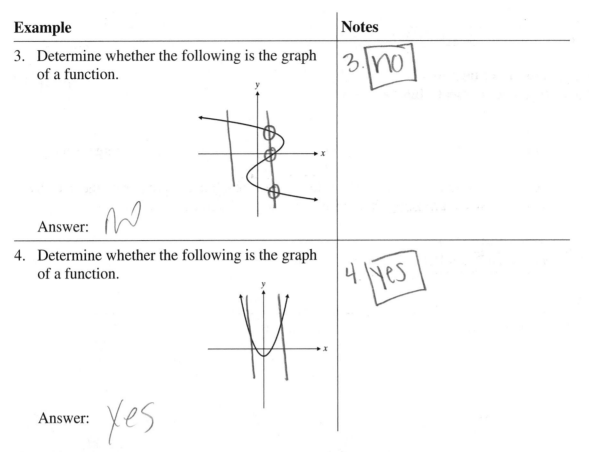

Answer: *(handwritten: no)*

3. no

---

4. Determine whether the following is the graph of a function.

Answer: *(handwritten: yes)*

4. Yes

Helpful Hints

A function cannot have two different ordered pairs with the same first coordinate. That is, each value of x must have one and only one value of y.

Concept Check

1. Explain why $y = -2$ is a function, but why $x = -2$ is not a function.

Practice

Determine whether each of the following is the graph of a function.

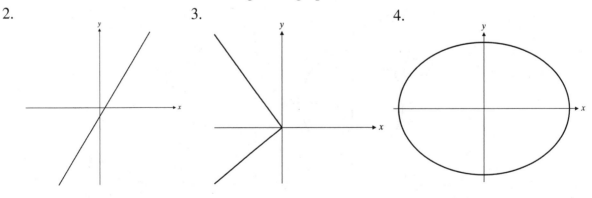

2.

3.

4.

Name: Sydney   Date: 2/8/16

Instructor: Schnabel   Section: _____

Introduction to Functions
Topic 19.3 Function Notation

Vocabulary

function notation   •   domain   •   range

1. If the name of a function is f and the variable is x, the function can be represented by
   the _function notation_ $f(x)$.

Step-by-Step Video Notes

Watch the Step-by-Step Video lesson and complete the examples below.

Example	Notes
1 & 2. Use function notation to rewrite the following functions using the given function names. $y = 9x - 2$, function name f $f(y) = 9x - 2$ $y = -16t^2 + 10$, function name h $h(t) = -16t^2 + 10$	1. $y = 9x \cdot 2$; function name f $f(x) = 9x - 2$ 2. $y = -16t^2 + 10$; function name h $h(t) = -16t^2 + 10$
3. Determine the domain and range of the function. $f(x) = -4x + 10$	$\{(2,2), (1,6), (0,10)(-1, 14), (-2, 18)\}$

X | 2 | 1 | 0 | 4 | 2
Y | 2 | 6 | 10 | 14 | 18

X:D | 2 | 1 | 0 | -1 | -2
Y:R | 2 | 6 | 10 | 14 | 18

Answer:

Example	**Notes**

4 & 5. Determine the domain and range of each function.

$$g(x) = x^2 - 4$$

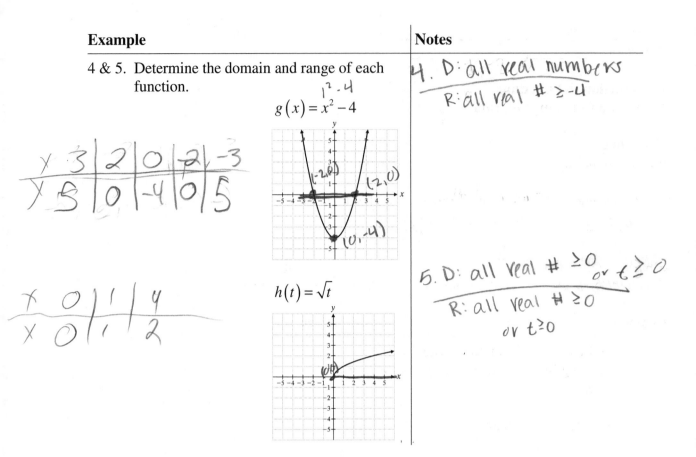

$1^2 \cdot 4$

x | 3 | 2 | 0 | 2 | -3

x | 5 | 0 | -4 | 0 | 5

$h(t) = \sqrt{t}$

x | 0 | 1 | 4

x | 0 | 1 | 2

4. D: all real numbers

R: all real # ≥ -4

5. D: all real # ≥ 0 or t ≥ 0

R: all real # ≥ 0 or t ≥ 0

Helpful Hints

Function notation is useful because the variable that makes up the domain is easily identified as the variable inside the parentheses. Read $f(x)$ as "f of x." It does not mean f times x.

Concept Check

1. A function g is defined as $x + y = 3$. Rewrite using the function notation $g(x)$.

Practice

Use function notation to rewrite the functions using the given function names.

2. $y = 3x - 7$, function name f

3. $y = -3x^2 + 5$, function name g

Determine the domain and range of the function based on its graph

4. $f(x) = \dfrac{4}{3}x - 3$

Introduction to Functions
Topic 19.4 Evaluating Functions

Vocabulary
function notation   •   evaluating a function   •   coordinate

1. When _evaluating a function_ at a certain value, substitute that value for the variable in the expression and simplify.

Step-by-Step Video Notes
Watch the Step-by-Step Video lesson and complete the examples below.

Example	Notes
1 & 2. If $f(x) = x + 8$, find the following. $f(2)$ 10 $f(-6)$ 2	1. $f(2) = 2 + 8$ $f(2) = 10$ $f(x) = x + 8$ 2. $f(-6) = -6 + 8$ $f(-6) = 2$
3–5. If $f(x) = 2x^2 - 4$, find each of the following. $f(5)$ 46 $f(-3)$ 14 $f(0)$ -4	$f(x) = 2x^2 - 4$ 3. $f(5) = 2(5)^2 - 4$ $2 \cdot 25 - 4$ $50 - 4$ $f(5) = 46$ 4. $f(-3) = 2(-3)^2 \cdot 4$ $2 \cdot 9 - 4$ $18 - 4$ $f(-3) = 14$ 5. $f(0) = 2(0)^2 \cdot 4$ $2 \cdot 0 - 4$ $f(0) = -4$

Example	Notes
6. The approximate length of a man's femur (thigh bone) is given by the function $f(x) = 0.5x - 17$, where x is the height of the man in inches. Find the approximate length of the femur of a man who is 70 inches tall. Substitute 70 for x. Simplify. $\begin{array}{r} 35 \\ -17 \\ \hline 18 \end{array}$	$f(x) = 0.5x - 17$ $f(70)$ $f(70) = 0.5(70) - 17$ $\quad\quad\quad 35 - 17$ $\boxed{f(70) = 18}$

Answer: 18

Helpful Hints

To find a function's value when x is some number a, we write $f(a)$. This point is shown on a graph by the coordinate $(a, f(a))$.

When evaluating a function, it is helpful to place parentheses around the value that is being substituted for x.

When a function defines a real-world relationship, be sure to use the correct units.

Concept Check

1. If $f(x) = (x-2)^2 - 9$, is $f(2) = f(-2)$? Is $f(5) = f(-1)$?

Practice

If $f(x) = 3x - 7$, find the following.

2. $f(-2)$

3. $f(2.6)$

If $h(t) = -16t^2 + 500$, find the following.

4. $h(3)$

5. $h(5)$

Introduction to Functions
Topic 19.5 Piecewise Functions

Vocabulary
independent variable   •   piecewise function   •   absolute value

1. A(n) _____ is a function whose definition changes depending on the
   value of the independent variable.

Step-by-Step Video Notes
Watch the Step-by-Step Video lesson and complete the examples below.

Example	Notes
1. If $f(x) = \begin{cases} 2x & \text{if} & x < 2 \\ x^2 & \text{if} & x \geq 2 \end{cases}$, find the following. $f(3)$ Determine which domain the value fits into, then substitute 3 for x in the corresponding expression and simplify. Answer:	
2. If $f(x) = \begin{cases} 2x+3 & \text{if } x \leq 0 \\ -x-1 & \text{if } x > 0 \end{cases}$, find the following. $f(2)$ $f(-6)$ $f(0)$	

Example	Notes
3. Susan's weekly salary (in dollars) is given by $$f(x) = \begin{cases} 10x & \text{if } 0 \le x \le 40 \\ 20x - 400 & \text{if } x > 40 \end{cases}, \text{ where } x \text{ is}$$ the number of hours worked per week. Find how much money Susan will make if she works the following number of hours in a week. 30 hours 60 hours	
4. Write the absolute value function $f(x) = |x|$ as a piecewise function.	

Helpful Hints

Each x value of a function must have one and only one y value. Do not substitute the value of x into more than one expression when evaluating a function.

Concept Check

1. Is $f(x) = \begin{cases} x^2 & \text{if } x < 0 \\ 4x^2 - 3x^2 & \text{if } x \ge 0 \end{cases}$ a piecewise function? Explain.

Practice

Find the value of $f(2)$ for each of the following piecewise functions.

2. $f(x) = \begin{cases} 5x - 7 & \text{if } x \le 2 \\ -x + 4 & \text{if } x > 2 \end{cases}$
 3. $f(x) = \begin{cases} x^2 + 6 & \text{if } x \le 0 \\ -x^2 - 1 & \text{if } x > 0 \end{cases}$
 4. $f(x) = \begin{cases} 0.5x & \text{if } x > 0 \\ x^{16} & \text{if } x \le 0 \end{cases}$

Name: Sydney

Instructor: Schnabel

Date: 2/10/16

Section: _____

Solving Systems of Linear Equations
Topic 20.1 Introduction to Systems of Linear Equations

Vocabulary
system of linear equations • ordered pair • solution to a system of linear equations

1. A _System of linear equations_ is a set of two or more linear equations containing the same variables.

Step-by-Step Video Notes
Watch the Step-by-Step Video lesson and complete the examples below.

Example	Notes
1. Determine whether $(3,-2)$ is a solution to the following system of equations. $x+3y=-3$ $4x+3y=6$ Answer: Yes	$3+3(-2) = -3$ $3-6 = -3$ $-3 = -3$ Yes $4(3)+3(-2) = 6$ $12-6 = 6$ $6 = 6$ Yes ✓
2. Determine whether $(4,3)$ is a solution to the following system. $7x-4y=16$ $5x+2y=24$ Answer: no	$7(4)-4(3) = 16$ $28-12 = 16$ $16 = 16$ Yes $5(4)+2(3) = 24$ $20+6 = 24$ $26 \neq 24$ no ✗

Example	Notes

Example

4. Determine whether $\left(\dfrac{4}{3}, \dfrac{1}{6}\right)$ is a solution to the following system.

$$2y = 1 - \dfrac{1}{2}x \checkmark$$
$$3x = 2 + 12y \checkmark$$

Answer: Y(s)

Notes

$2\left(\dfrac{1}{6}\right) = 1 - \dfrac{1}{2}\left(\dfrac{4}{3}\right)$

$\dfrac{2}{1}\left(\dfrac{1}{6}\right) = 1 - \dfrac{1}{2}\left(\dfrac{4}{3}\right)$

$\dfrac{2}{6} = \dfrac{6}{6} - \dfrac{4}{6}$

$\dfrac{2}{6} = \dfrac{2}{6}$

$\boxed{\text{YES}}$ ✓

$3\left(\dfrac{4}{3}\right) = 2 + 12\left(\dfrac{1}{6}\right)$

$\dfrac{9}{3}\left(\dfrac{4}{3}\right) = 2 + \dfrac{72}{6}\left(\dfrac{1}{6}\right)$

$\dfrac{36}{9} = 2\dfrac{72}{36} \dfrac{72}{36}$

$\dfrac{36}{9} : \dfrac{144}{36}$

$4 = 4$

$\boxed{\text{YES}}$

Helpful Hints

If the point (x, y) exists on two lines, then it is a solution to the system of the equations which contains those lines.

Concept Check

1. If $(2,1)$ is the solution to a system of two equations and a third equation is added to the system, is $(2,1)$ still a solution? Explain.

Practice

Determine whether $(4, -1)$ is a solution to the following systems.

2. $x + 2y = 2$
   $5x + 3y = 16$

3. $3x - 4y = 16$
   $5x + 6y = 14$

Determine whether $(0.5, 3)$ is a solution to the following systems.

4. $2x = 3y - 1$
   $\dfrac{1}{3}y = 6x - 2$

5. $0.6x + 0.4y = 1.5$
   $8x + 3y = 12$

Name: _Sydney_   Date: _2/10/16_

Instructor: _Schnabel_   Section: _____

Solving Systems of Linear Equations
Topic 20.2 Solving by the Graphing Method

Vocabulary

inconsistent system of equations • dependent system of equations • point of intersection

1. A(n) _inconsistent system_ has no solution. Its graph will be two parallel lines, which do not intersect. There are no ordered pairs in common.

2. A(n) _dependent system_ has an infinite number of solutions. This means that the graphs of the two equations will show the same line.

Step-by-Step Video Notes

Watch the Step-by-Step Video lesson and complete the examples below.

Example	Notes
1. Find the solution (the point of intersection of the two lines). 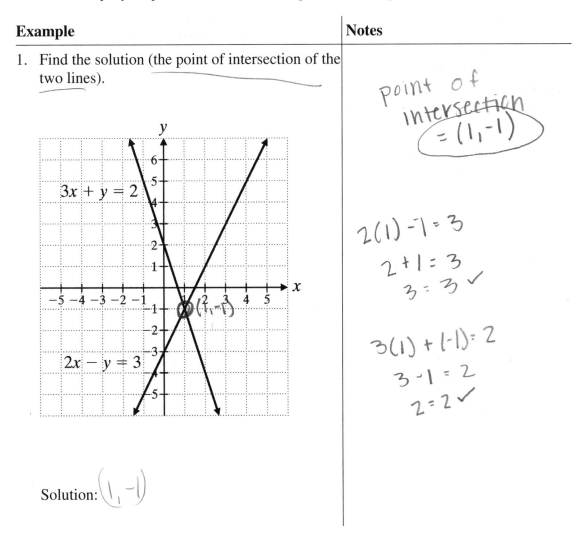 Solution: $(1, -1)$	point of intersection $= (1, -1)$ $2(1) - 1 = 3$ $2 + 1 = 3$ $3 = 3$ ✓ $3(1) + (-1) = 2$ $3 - 1 = 2$ $2 = 2$ ✓

247
Copyright © 2012 Pearson Education, Inc.

Example	Notes

3. Find the solution to the system of equations by graphing.

$$y = x - 1$$

$$2x + y = 8$$

$x = -2x + 8$

Answer:

$(3, 2)$

$2 = 3 - 1 \checkmark$

$6 + 2 = 8 \checkmark$

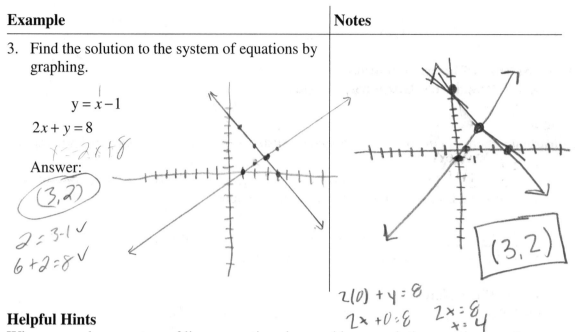

$(3, 2)$

$2(0) + y = 8$

$2x + 0 = 8 \quad 2x = 8$

$x = 4$

Helpful Hints

When you solve a system of linear equations by graphing, graph the equations on the same graph. Find the solution, which is the point of intersection of the lines. Write the solution as an ordered pair. Verify your solution by substituting the ordered pair into both equations.

Concept Check

1. What are the three possible types of solutions to a system of linear equations?

Practice

Find the solution to the system of equations by graphing.

2. $y = x + 1$
    $y = 3x - 1$

3. $x + y = 5$
    $y = \dfrac{1}{2}x - 1$

4. $2x + 3y = 9$
    $y = -2x + 3$

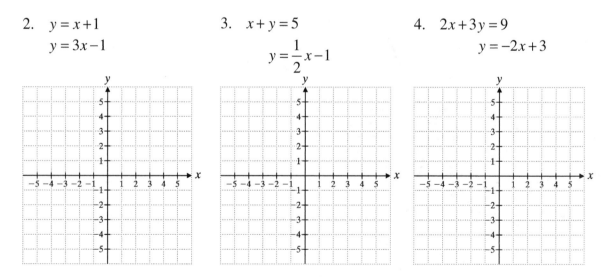

Name: _Sydney_ _Schnabel_   Date: _2/10/16_
Instructor: _Schnabel_   Section: _____

_US_

Solving Systems of Linear Equations
Topic 20.3 Solving by the Substitution Method

Vocabulary
substitution method   •   system of equations   •   no solution

1. The _substitution method_ involves choosing either equation in a system and solving for either variable and substituting the result into the other equation to solve for the remaining variable.

Step-by-Step Video Notes
Watch the Step-by-Step Video lesson and complete the examples below.

Example	Notes
1. Solve the system of equations by substitution. $4x + 3y = 50$ $\quad$ $4y+6x=50$ $\qquad y = 2x$ $\quad$ $10y=50$ $\qquad\qquad y=5$ Substitute $2x$ for y in the first equation. Simplify and solve for x. Use this value for x in one equation to find y. Check your solution in the other equation. Solution: $(5, 10)$	$4x + 3(2x)=50$ $4x + 6x = 50$ $\qquad 10x = 50$ $\qquad\boxed{x = 5}$ $y = 2(5)$ $\boxed{y = 10}$ $\boxed{(5, 10)}$
2. Solve the system of equations by substitution. $\qquad y = x - 5$ $3x - 2y = 17$ $3x - 2(x-5) = 17$ $3x - 2x - 10 = 17$ $\qquad\quad +10 \quad +10$ $\qquad\qquad\qquad x = 27$ Solution: $(27, 22)$	$3x - 2(x \cdot 5) = 17$ $3x - 2x - 10 = 17$ $\qquad x - 10 = 17$ $\qquad\boxed{x = 27}$ $y = 27 - 5$ $\boxed{y = 22}$

249
Copyright © 2012 Pearson Education, Inc.

Example	Notes
4. Solve the system of equations by substitution. Solve one equation for one variable.	$4x - 24y = 20$
$4x - 24y = 40$ $24y + 40 - 24y = 40$ 40	$x - 6y = 10$ *many solutions*
$x - 6y = 10$ -40	$x = 6y + 10$
$x = 6y + 10$ $0 = 0$	$x = 6y + 10$
$4(6y + 10) - 24y = 40$	$4(6y + 10) - 24y = 40$
	$24y + 40 - 24y = 40$
Solution: many solutions	$40 = 40$
5. Solve the system of equations by substitution.	$x + 3y = 5$
$x + 3y = 5$ $x = 5 - 3y$	$x = -3y + 5$
$4x + 12y = 40$ $4(5 - 3y) + 12y = 40$	$4(-3y + 5) + 12y = 40$
$20 - 12y + 12y = 40$	$-12y + 20 + 12y = 40$
$20 = 40$	$20 \neq 40X$
Solution: no solution	*no solution*

Helpful Hints

If one of the equations has a variable with a coefficient of 1 or -1, choose that equation to solve. This will almost always make this step easier.

Concept Check

1. Describe a situation where solving a system by substitution would be easier than solving it by graphing.

Practice

Solve the system of equations by substitution.

2. $4x + 3y = 38$
     $y = 5x$

4. $3x + 4y = 8$
     $2x + y = -3$

3. $6x + 5y = 10$
     $x = y + 9$

5. $y = 4x + 7$
     $8x - 2y = 19$

Name: Sydney Schnabel

Date: 2/11/16

Instructor: Schnabel

Section: _____

Solving Systems of Linear Equations
Topic 20.4 Solving by the Elimination Method

CS

Vocabulary

Elimination Method   •   many solutions   •   no solution

1. When solving systems of equations, you can use the _Elimination method_ to solve the system if the coefficients of either variable in the equations are opposites.

Step-by-Step Video Notes

Watch the Step-by-Step Video lesson and complete the examples below.

Example	Notes
1. Solve the system of equations. $$7x - 3y = 1$$ $$5x + 3y = 11$$ $7-3y=1$ $-7 \quad -7$ $-3y=6$ $y=2$ $12x + 6y = 12$ $x=1$ Solution: $(1,2)$	$7x - 3y = 1$ $(+)5x + 3y = 11$ $12x = 12$ $\boxed{x=1}$ $7(1) - 3y = 1$ $7 - 3y = 6$ $-3y = -6$ $\boxed{y=2}$ $\boxed{(1, 2)}$
2. Solve the system of equations by the Elimination Method. $$3x + 7y = 22$$ $$2x - 7y = 3$$ $$5x = 25$$ Check to see if the coefficients of either variable are opposites. $x = 5$ $15 + 7y = 22$ $-5 \qquad -5$ $7y = 7$ $y = 1$ Solution: $(5, 1)$	$3x + 7y = 22$ $(+)2x - 7y = 3$ $5x = 25$ $\boxed{x = 5}$ $\boxed{(5, 1)}$ $3(5) + 7y = 22$ $15 + 7y = 22$ $7y = 7$ $\boxed{y = 1}$

Copyright © 2012 Pearson Education, Inc.

Example	Notes

4. Solve the system of equations by the Elimination Method.

$$4x + 3y = -5$$
$$7x + 2y = 14$$

Handwritten left:
$$7(4y + 3y = -5)$$
$$-4(7x + 2y = 14)$$

$$7y1 - 14 = 14$$
$$+14 \quad +4$$
$$x = 4$$

Solution: $(4, -7)$

Handwritten notes column:
$$2(4x + 3y = -5)$$
$$3(7x + 2y = 14)$$

$$8x + 6y = -10$$
$$(-)21x + 6y = 42$$
$$\overline{}$$
$$-13x = -52$$
$$\boxed{x = 4}$$

$$4(4) + 3y = -5$$
$$16 + 3y = -5$$
$$3y = -21$$
$$\boxed{y = -7}$$

$$(4, -7)$$

5. Solve the system of equations by the Elimination Method.

$$3x + 2y = 18$$
$$\underline{-3x - 2y = 14}$$
$$0 = 32$$

Solution: No Solution

Handwritten notes column:
$$3x + 2y = 18$$
$$(-)-3x - 2y = 14$$
$$\overline{}$$
$$0 \neq 32$$

$$\boxed{\text{no solution}}$$

Helpful Hints
An easy way to use elimination is to pick which variable you want to eliminate and multiply each equation by the coefficient of that variable term in the other equation. It may be necessary to also multiply by a negative number to produce opposite variable terms.

Concept Check
1. Describe a situation where solving a system by elimination would be easier than solving by substitution.

Practice
Solve the system by the Elimination Method.

2. $7x + 4y = 8$
   $-7x - 6y = 2$

4. $7x + 5y = 98$
   $8x - 2y = 58$

3. $5x - 6y = 17$
   $2x - 4y = 6$

5. $48x - 33y = 57$
   $-16x + 11y = 19$

Introduction to Polynomials and Exponent Rules
Topic 21.1 Introduction to Polynomials

Vocabulary

term • coefficient • like terms • polynomial • descending order • monomial
degree of a term • degree of a polynomial • binomial • trinomial • exponent

1. The _degree of a term_ is the highest exponent of the base in that term. If there is more than one variable, it is the sum of the exponents of all of the variables in the term.

2. A _Polynomial_ in the variable x is the sum of a finite number of terms of the form ax^n, where a is any real number and n is a whole number.

Step-by-Step Video Notes
Watch the Step-by-Step Video lesson and complete the examples below.

Example	Notes
2. Find the degree of the term $4ab^2$. The degree of a term is the highest exponent of the base in that term. This term has two variables, __a__ and __b__. The exponent of __a__ is ☐1, and of __b__ is ☐. Answer: 3	$4ab^2$ $4a^{1}b^{2}$ degree = 3
4. For the polynomial $5x^3 + 8x^2 - 20x - 2$, find the degree of each term. Then find the degree of the polynomial. The degree of the first term is ☐3, the degree of the second term is ☐2, the degree of the third term is ☐1, and the degree of the last term is ☐ since it is a constant. Answer: 7	$5x^3 + 8x^2 - 20x - 2$ $5x^3$: degree ③ ✗ $8x^2$: degree 2 $20x$: degree 1 2: degree 0 degree of polynomial = degree 3

Example	Notes
6. State the degree of the polynomial $5x + 3x^3$, and whether it is a monomial, a binomial, or a trinomial.	$5x + 3x^3$ $= $ degree 3 * binomial
Answer: 4	
9. Evaluate the polynomial $P(x) = 2x^2 - 6x + 7$ at $x = -2$. $2(4) - 6(-2) + 7$ $8 + 12 + 7$ Answer: 27	$2x^2 - 6x + 7$ $2(-2)^2 - 6(-2) + 7$ $2(4) + 12 + 7$ $8 + 12 + 7$ $P(-2) = 27$

Helpful Hints

You can combine like terms by adding or subtracting the coefficients of the terms.

A polynomial in x is said to be written in descending order if it is written so that the exponents on the variable x decrease from left to right.

The degree of a polynomial is the highest degree of any of its terms. The polynomial 0 is said to have no degree. A polynomial consisting of a constant only is said to have degree 0.

Concept Check

1. Is the polynomial $24x^3 + 8x^2 + 4x + 2$ of greater degree than the monomial $-2x^5$?

Practice

State the degree of the polynomial, and tell if it is a monomial, binomial, or trinomial.

2. $72x^3 + 45x^2 + 27x$

3. $15x^4y^2 - 22x^3y$

Evaluate the polynomial at the given value of the variable.

4. $f(x) = 2x^2 + 3x - 9$ for $x = 1.5$

5. $h(t) = -16t^2 + 400$ for $t = 4$

Name: _Sydney_ Date: _1/20/16_

Instructor: _Schnabel_ Section: _____

Introduction to Polynomials and Exponent Rules
Topic 21.2 Addition of Polynomials

Vocabulary
adding polynomials • coefficients • like terms

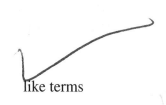

1. When _adding polynomials_, combine like terms.

Step-by-Step Video Notes
Watch the Step-by-Step Video lesson and complete the examples below.

Example	Notes
2. Add. $(5x^2 - 6x - 12) + (-3x^2 - 9x + 5)$ Remove parentheses and identify like terms. $5x^2 - 6x - 12 - 3x^2 - 9x + 5$ Combine like terms. $2x^2 - 15x - 7$ Answer: $2x^2 - 15x - 7$	$(5x^2 - 6x - 12) + (-3x^2 - 9x + 5)$ $5x^2 - 3x^2 = 2x^2$ $-6x - 9x = -15x$ $-12 + 5 = -7$ $2x^2 - 15x - 7$
3. Add. $(7x^2 + 8x + 9) + (13x^2 - 10x + 5)$ Remove parentheses and identify like terms. Combine like terms. We can also use a vertical format to help visualize the addition. $7x^2 + \ \ 8x + 9$ $+13x^2 - 10x + 5$ $20x^2 - 2x \ \ 4$ Answer: $20x^2 - 2x + 4$	$(7x^2 + 8x + 9) + (13x^2 - 10x + 5)$ $7x^2 + 8x + 9$ $+13x^2 - 10x + 5$ $20x^2 - 2x + 14$

Example	Notes

Example

4. Add.

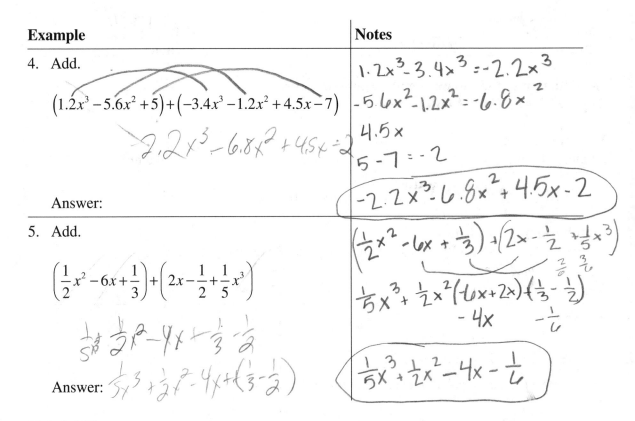

$(1.2x^3 - 5.6x^2 + 5) + (-3.4x^3 - 1.2x^2 + 4.5x - 7)$

$2.2x^3 - 6.8x^2 + 4.5x - 2$

Answer:

Notes

$1.2x^3 - 3.4x^3 = -2.2x^3$

$-5.6x^2 - 1.2x^2 = -6.8x^2$

$4.5x$

$5 - 7 = -2$

$\boxed{-2.2x^3 - 6.8x^2 + 4.5x - 2}$

5. Add.

$\left(\dfrac{1}{2}x^2 - 6x + \dfrac{1}{3}\right) + \left(2x - \dfrac{1}{2} + \dfrac{1}{5}x^3\right)$

$\dfrac{1}{5}x^3 \quad \dfrac{1}{2}x^2 - 4x + \dfrac{1}{3} - \dfrac{1}{2}$

Answer: $\dfrac{1}{5}x^3 + \dfrac{1}{2}x^2 - 4x + \left(\dfrac{1}{3} - \dfrac{1}{2}\right)$

$\left(\dfrac{1}{2}x^2 - 6x + \dfrac{1}{3}\right) + \left(2x - \dfrac{1}{2} + \dfrac{1}{5}x^3\right)$

$\dfrac{1}{5}x^3 + \dfrac{1}{2}x^2 \, (-6x + 2x) + \left(\dfrac{1}{3} - \dfrac{1}{2}\right)$

$-4x \qquad -\dfrac{1}{6}$

$\boxed{\dfrac{1}{5}x^3 + \dfrac{1}{2}x^2 - 4x - \dfrac{1}{6}}$

Helpful Hints

You can add polynomials by combining like terms. While each polynomial may not be given in descending order, arrange the sum in descending order.

If you use a vertical format to add polynomials, be sure each column contains terms of the same degree.

Concept Check

1. How many terms are in the polynomial sum $(4x^3 - 7x^2 + 3x) + (-2x^2 + 9x + 3)$?

Practice

Add.

2. $(-4x^3 + 3x^2 - 2) + (7x^3 - 8x^2 - 3)$

4. $\left(-\dfrac{1}{2}x^2 - 5x - 6\right) + \left(8x + \dfrac{3}{4}x^2 + \dfrac{2}{3}\right)$

3. $(4.4x^3 - 0.13x^2 + 2.11x) + (-0.07x^2 - 1.89x + 6)$

5. $(13x^2 + 2 - 5x) + (13x - 9x^2 + 7)$

Introduction to Polynomials and Exponent Rules
Topic 21.3 Subtraction of Polynomials

Vocabulary
Adding polynomials　　•　　subtracting polynomials　　•　　like terms

1. When _Subtracting polynomials_, change the sign of each term in the second polynomial and add the result to the first polynomial by combining like terms.

Step-by-Step Video Notes
Watch the Step-by-Step Video lesson and complete the examples below.

Example	Notes
1. Subtract. $(2x+3)-(x-5)$ Change the sign of each term in the second polynomial and add. $2x+3-\boxed{x}+\boxed{5}$ Add the polynomials by combining like terms. $2x-\boxed{x}+3+\boxed{5}$ Answer: $X+8$	$(2x+3)+(-x+5)$ $2x+3-1x+5$ $2x-1x = 1x$ $3+5=8$ $\boxed{x+8}$
2. Subtract. $(-2x^3+7x^2-3x-1)-(-6x^3-9x^2-x+4)$ Change the sign of each term in the second polynomial and add. You can also use a vertical format to help visualize the addition. $\begin{array}{l}-2x^3+7x^2-3x-1\\+(-6x^3-9x^2+x+4)\\\hline +4x^3+16x^2-2x-5\end{array}$ $\rightarrow$ $\begin{array}{l}-2x^3+7x^2-3x-1\\+\boxed{}\end{array}$ Answer: $4x^3+16x^2-2x-5$	$(-2x^3+7x^2-3x-1)+(6x^3+9x^2+x-4)$ $-2x^3+7x^2-3x-1$ $+6x^3+9x^2+x-4$ $\boxed{4x^3+16x^2-2x-5}$

Example	Notes
3. Subtract. $(-3x^4 + 5x^2 + 2) - (6x^3 - 10x^2 + 2x - 1)$ Answer:	$(-3x^4 + 5x^2 + 2) + (-6x^3 + 10x^2 - 2x + 1)$ $-3x^4 - 6x^3 + 15x^2 - 2x + 3$
5. Subtract. $(-6x^2y - 3xy + 7xy^2) - (5x^2y - 8xy - 15x^2y^2)$ Answer:	$(-6x^2y - 3xy + 7xy^2) + (-5x^2y + 8xy +$ $15x^2y^2 - 6x^2y - 5x^2y + 7xy^2 - 3xy + 8$ $15x^2y^2 - 11x^2y + 7xy^2 + 5xy$

Helpful Hints

Use extra care in determining which terms are like terms when polynomials contain more than one variable. Every exponent of every variable in the two terms must be the same if the terms are to be like terms.

If you use a vertical format to subtract polynomials, be sure to change the sign of each term in the second polynomial before adding.

Concept Check

1. How many terms are in the difference $(4xy - 7x^2y^2 + 3y) - (-2x^2y^2 + 9xy + 3)$?

Practice

Subtract.

2. $(4x^2 + 6x - 9) - (-5x^2 - 6x + 2)$ 　　　　4. $(6.3x^2 - 1.8x + 3.5) - (3.2x - 0.7x^2 - 4.9)$

3. $(15x^5 - 8x^3 + 5x - 3) - (-5x^5 + 12x^4 + 6x^2)$ 　5. $(a^4 - 7ab + 3ab^2 - 2b^3) - (2a^4 + 4ab - 6b^3)$

Name: Sydney Schnabel

Date: 4/21/16

Instructor: Schnabel

Section: _____

Introduction to Polynomials and Exponent Rules
Topic 21.4 Product Rule for Exponents

Vocabulary

exponential expression   •   product rule for exponents   •   base

1. A variable or a number raised to an exponent is a(n) _exponential expression_

2. The _product rule for exponents_ states that to multiply two exponential expressions that have like bases, keep the base and add the exponents, or $x^a \cdot x^b = x^{a+b}$.

Step-by-Step Video Notes

Watch the Step-by-Step Video lesson and complete the examples below.

Example	Notes
3. Multiply $x^3 \cdot x \cdot x^6$. The base of each factor is $\boxed{\text{x}}$. Keep this base and add the exponents. Note that even though it is not written, the exponent of the term x is $\boxed{\text{x}}$. $x^3 \cdot x \cdot x^6 = x^{\left(\boxed{3}+\boxed{1}+\boxed{6}\right)}$ Answer:	$x^3 \cdot x \cdot x^6 \quad (x \cdot x \cdot x)\cdot(x)\cdot(x \cdot x \cdot x \cdot x \cdot x \cdot x)$ x^{3+1+6} $= x^{10}$ x^{3+1+6}
5. Simplify $2^3 \cdot 2^5$, if possible. Write your answer using exponential notation. 2^8 Answer:	$2^3 \cdot 2^5$ $2^{3+5} = \boxed{2^8}$

Example	Notes
7. Multiply $(3a)^2 \cdot (3a)^4$. The base of each factor is 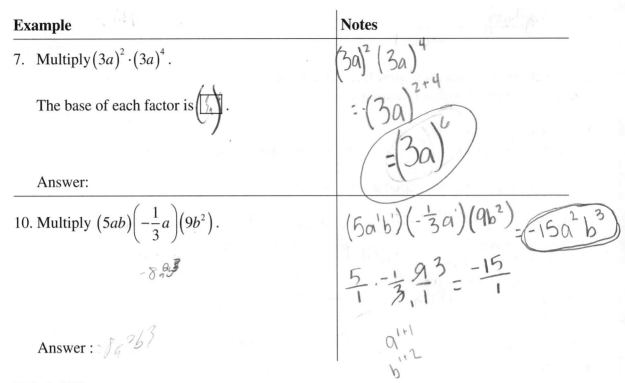. Answer:	$(3a)^2 (3a)^4$ $= (3a)^{2+4}$ $= (3a)^6$
10. Multiply $(5ab)\left(-\dfrac{1}{3}a\right)(9b^2)$. $-8a^3$ Answer: $8a^2b^3$	$(5a'b')\left(-\tfrac{1}{3}a'\right)(9b^2) = -15a^2b^3$ $\dfrac{5}{1} \cdot \dfrac{-1}{3} \cdot \dfrac{9^3}{1} = \dfrac{-15}{1}$ a^{1+1} b^{1+2}

Helpful Hints

It is important that you apply the product rule even when the exponent is 1. Every variable that does not have a written exponent is understood to have an exponent of 1.

To multiply exponential expressions with coefficients, first multiply the coefficients, and then multiply the variables with exponents separately.

Concept Check

1. Can you use the product rule for exponents to simplify $p^2 \cdot g^2$? Explain.

Practice

Multiply.

2. $w^{12} \cdot w$

4. $(4xy)\left(-\dfrac{1}{8}x^4 y^2\right)(4xy^4)$

3. $(-8x^4)(-5x^3)$

5. $(-6.5pq^5)(-p^3)(2p^4q)$

Name: Sydney

Instructor: Schnabel

Date: 1/21/16

Section: ___

Introduction to Polynomials and Exponent Rules
Topic 21.5 Power Rule for Exponents

Vocabulary

product rule for exponents • quotient raised to a power • exponential expression
power rule for exponents • product raised to a power • exponential notation

1. The _power rule for exponents_ states that $\left(x^a\right)^b = x^{ab}$.

2. The rule for a _quotient raised to a power_ is demonstrated by the equation $\left(\dfrac{x}{y}\right)^a = \dfrac{x^a}{y^a}$, if $y \neq 0$.

Step-by-Step Video Notes

Watch the Step-by-Step Video lesson and complete the examples below.

Example	Notes
1–3. Simplify. Write your answer using exponential notation. 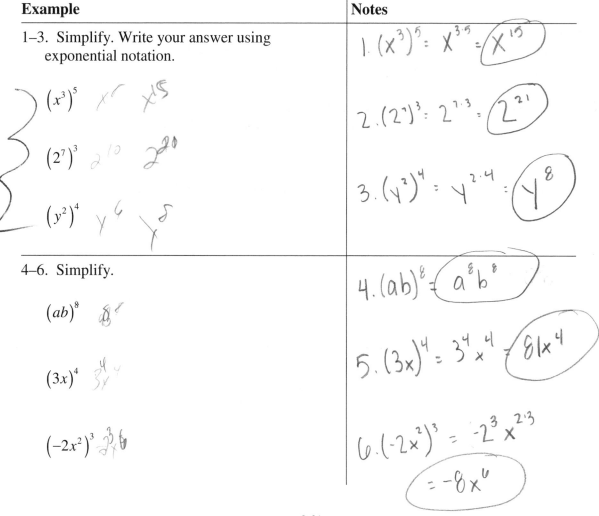	1. $\left(x^3\right)^5 = x^{3\cdot5} = \boxed{x^{15}}$ 2. $\left(2^7\right)^3 = 2^{7\cdot3} = \boxed{2^{21}}$ 3. $\left(y^2\right)^4 = y^{2\cdot4} = \boxed{y^8}$
4–6. Simplify. $(ab)^8$ $(3x)^4$ $\left(-2x^2\right)^3$	4. $(ab)^8 = \boxed{a^8 b^8}$ 5. $(3x)^4 = 3^4 x^4 = \boxed{81x^4}$ 6. $\left(-2x^2\right)^3 = -2^3 x^{2\cdot3}$ $= \boxed{-8x^6}$

Example	Notes

8. Simplify $\left(\dfrac{3}{w}\right)^4$. Write your answer using exponential notation.

Answer: $\dfrac{3^4}{w^4}$

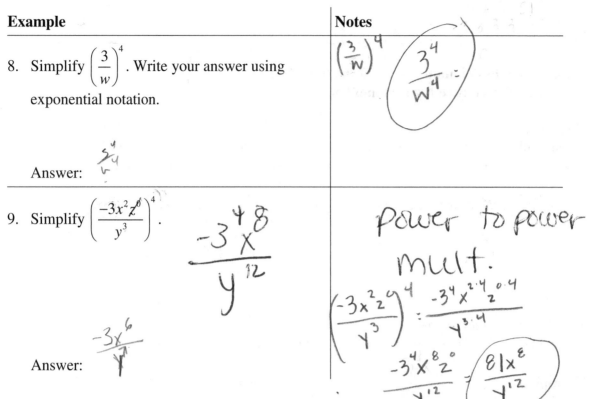

9. Simplify $\left(\dfrac{-3x^2z^0}{y^3}\right)^4$.

$\dfrac{-3^4x^8}{y^{12}}$

Answer: $\dfrac{-3x^6}{y}$

Notes (handwritten):

$\left(\dfrac{3}{w}\right)^4 \quad \dfrac{3^4}{w^4} =$

power to power mult.

$\left(\dfrac{-3x^2z^0}{y^3}\right)^4 = \dfrac{-3^4x^{2\cdot4}z^{0\cdot4}}{y^{3\cdot4}}$

$\dfrac{-3^4x^8z^0}{y^{12}} = \dfrac{81x^8}{y^{12}}$

Helpful Hints

If a product in parentheses is raised to a power, the parentheses indicate that each factor inside the parentheses must be raised to that power.

When simplifying expressions by using multiple rules involving exponents, be careful determining the correct sign, especially if there is a negative coefficient.

Concept Check

1. A student simplified $\left(-3x^2\right)^4$ as $-81x^8$. Is this correct? If not, what is the error?

Practice
Simplify.

2. $\left(d^4\right)^5$

3. $\left(-8x^4\right)^2$

4. $\left(\dfrac{p}{2}\right)^5$

5. $\left(-\dfrac{2y^0z^3}{3x^5}\right)^4$

Name: _Sydney_____   Date: _1/21/16___
Instructor: _____Schnabel_____   Section: _____

Multiplying Polynomials
Topic 22.1 Multiplying by a Monomial

Vocabulary
Distributive Property   •   product rule for exponents   •   monomial

1. A _monomial_ is a polynomial with exactly one term.

2. To multiply a monomial by a polynomial, use the _Distributive Property_ to multiply the monomial by each term in the polynomial.

Step-by-Step Video Notes
Watch the Step-by-Step Video lesson and complete the examples below.

Example	Notes
1. Multiply $3x^2(5x-2)$. $\qquad 15x^3 - 6x^2$ Answer:	$3x^2(5x-2)$ $(3x^2)(5x) - (3x^2)(2)$ $\boxed{15x^3 - 6x^2}$
2. Multiply $2x(x^2+3x-1)$. Multiply the monomial by each term in the parentheses. $2x^3 + 6x^2 - 2x$ Answer:	$2x(x^2+3x-1)$ $(2x)(x^2) + (2x)(3x) - (2x)(1) =$ $\boxed{2x^3 + 6x^2 - 2x}$

Example	Notes
3. Multiply $-6xy\left(x^3+2x^2y-y^2\right)$. $-6x^4y-12x^3y^2+6xy^3$ Answer:	$(-6xy)(x^3)+(-6xy)(2x^2y)-(-6xy)(y^2)$ $-6x^4y-12x^3y^2+6xy^3$
5. Multiply $\left(2x^2-3x+8\right)(-7x)$. Use the Commutative Property to write the monomial first. $-7x\left(2x^2-3x+8\right)$ Answer: $-14x^3+21x^2-56x$	$(-7x)(2x^2-3x+8)$ $(-7x)(2x^2)-(-7x)(3x)+(-7x)(8)$ $-14x^3-(-21x^2)+(-56x)$ $-14x^3+21x^2-56x$

Helpful Hints

Remember with a term such as $7x$ that the exponent on x is 1. When you multiply, add the powers, so $7x\cdot x^2 = 7x^{(1+2)} = 7x^3$.

Multiplication is commutative. The order in which terms are multiplied doesn't matter. A monomial can be after a polynomial, but a monomial is usually written before a polynomial.

Concept Check

1. Which multiplication would you perform first to simplify $(2x)(3y)\left(x^2+4y\right)$? Why?

Practice

Multiply.

2. $6x^3\left(-3x^2+2x\right)$

4. $-3xy^2\left(x^2+8xy-4y^2\right)$

3. $5x\left(x^2-4x-7\right)$

5. $\left(x^3-6x+12\right)(-3x)$

Name: Sydney

Instructor: Schnabel

Date: 1/22/16

Section: _____

Multiplying Polynomials
Topic 22.2 Multiplying Binomials

Vocabulary

binomial • FOIL method • polynomial

1. The _FOIL method_ for multiplying two binomials means multiply the first terms, the outer terms, the inner terms, and then the last terms.

2. A _binomial_ is a polynomial with exactly two terms.

Step-by-Step Video Notes
Watch the Step-by-Step Video lesson and complete the examples below.

Example	Notes
1. Multiply $(3x+1)(x+4)$.	$(3x+1)(x+4)$
Multiply each term in the first binomial by each term in the second binomial.	$3x^2 + 12x + 1x + 4$
$3x\left(\boxed{}\right) + 3x\left(\boxed{}\right) + 1\left(\boxed{}\right) + 1\left(\boxed{}\right)$	$\boxed{3x^2 + 13x + 4}$
Combine like terms to simplify the expression.	
$3x^2 + 12x + x + 4$ $3x^2 + 13x + 4$	
Answer: $3x^2 + 13x + 4$	
3. Multiply $(4x-9y)(8x-3)$.	$(4x-9y)(8x-3)$
$32x^2 - 12x - 72xy + 27y$	$32x^2 - 12x - 72xy + 27y$
	$\boxed{32x^2 - 72xy - 12x + 27y}$
Answer: $32x^2 - 12x - 72xy + 27y$	

Example	Notes
4. Multiply $(x+3)(x+5)$ using the FOIL method.	$(x+3)(x+5)$
Multiply the first terms.	$x^2+5x+3x+15$
Multiply the outer terms.	$=\boxed{x^2+8x+15}$
Multiply the inner terms.	
Multiply the last terms.	
Answer: $x^2+8x+15$	
6. Multiply $(-2+7x)(-9x+5)$ using the FOIL method.	$(-2+7x)(-9x+5)$
	$18x-10-63x^2+35x$
	$=\boxed{-63x^2+53x-10}$
Answer: $-63x^2+53x-10$	

Helpful Hints

The FOIL method is just a way to help you remember how to multiply binomials. It is only used to multiply a binomial by a binomial.

Concept Check

1. Would you use the FOIL method when multiplying $(7x^2-3x^2)(-4x^3+18y)$? Why or why not?

Practice

Multiply.

2. $(x+4)(3x+7)$

3. $(5x-12y)(9x-4)$

Multiply using the FOIL method.

4. $(x^2+3)(x^2+8)$

5. $(4x-6)(-3x-2)$

Multiplying Polynomials
Topic 22.3 Multiplying Polynomials

Vocabulary
polynomial • multiplying polynomials • Distributive Property

1. When __multiplying polynomials__, multiply each term in the first polynomial by every term in the second polynomial. Then write the sum of the products in descending order.

Step-by-Step Video Notes
Watch the Step-by-Step Video lesson and complete the examples below.

Example	Notes
1. Multiply $(x+4)(3x^2+x+2)$. Multiply each term in the first polynomial by every term in the second polynomial. Combine like terms. Write the final polynomial in descending order. Answer: $3x^3 + 13x^2 + 6x + 8$	$(x+4)(3x^2+x+2)$ $(x)(3x^2)+(x)(x)+(x)(2)+4(3x^2)+(4)(x)+(4)(2)$ $3x^3+x^2+2x+12x^2+4x+8$ $\boxed{3x^3+13x^2+6x+8}$
2. Multiply $(3x-1)(6x^2-5x+8)$. $18x^3-15x^2+24x-6x^2+5x-8$ Answer: $18x^3-21x^2+29x-8$	$(3x-1)(6x^2-5x+8)$ $(3x)(6x^2)-(3x)(5x)+(3x)(8)$ $-1(6x^2)-(-1)(5x)+(-1)(8)$ $18x^3-15x^2+24x-6x^2+5x-8$ $\boxed{18x^3-21x^2+29x-8}$

Example	Notes

4. Multiply $(4x^2 + 9x + 7)(2x^2 - 6x - 5)$.

$$8x^4 - 6x^3 - 60x^2 - 87x - 35$$

Answer: $4x^4 - 6x^3 - 58x^2 - 87x$

$(4x^2)(2x^2) - (4x^2)(6x) - (4x^2)(5)$

$(9x)(2x^2) - (9x)(6x) - (9x)(5)$

$(7)(2x^2) - (7)(6x) - (7)(5)$

$8x^4 - 24x^3 - 20x^2 + 18x^3 - 54x^2 - 45x +$

---

5. Multiply $(x+1)(x+2)(x+3)$.

Multiply the first two polynomials, then multiply the product by the third polynomial.

$(x+3)(x^2 + 3x + 2)$

$x^3 + 4x + 2x + 3x^2 + 6x + 6$

Answer: $x^3 + 3x^2 + 12x + 6$

$(x+1)(x+2)$

$x^2 + 2x + 1x + 2 = x^2 + 3x + 2$

$(x+3)(x^2 + 3x + 2)$

$x^3 + 3x^2 + 2x + 3x^2 + 9x + 6$

$$x^3 + 6x^2 + 11x + 6$$

Helpful Hints
A good way to keep terms organized when multiplying bigger polynomials is to multiply the polynomials vertically. This is like multiplying multi-digit numbers vertically.

You can also multiply three or more polynomials; just multiply them two at a time.

Concept Check
1. Which multiplication would you perform first to simplify $(x+4)(x^2 + 7 - 3x)(x+1)$? Why?

Practice
Multiply.

2. $(8x+4)(x^2 + 7x - 6)$

4. $(x^2 + 6x + 8)(2x^2 - x - 6)$

3. $(5x^2 - 9x + 4)(6x - 2)$

5. $(4x - 2)(3x + 7)(x + 5)$

Multiplying Polynomials

Topic 22.4 Multiplying the Sum and Difference of Two Terms

Vocabulary

FOIL method   •   multiplying binomials: a sum and a difference

1. The rule _multiplying binomials: a sum and a difference_ states that $(a+b)(a-b)=a^2-b^2$, where a and

   b are numbers or algebraic expressions.

Step-by-Step Video Notes

Watch the Step-by-Step Video lesson and complete the examples below.

Example	Notes
1. Multiply $(5x+4)(5x-4)$. Multiply each term in the first polynomial by each term in the second polynomial. $5x^2-16$ $25x^2-16$ Answer:	$(5x+4)(5x-4)$ $25x^2-20x+20x-16$ $\boxed{25x^2-16}$
2. Multiply $(x+5)(x-5)$. $25x^2$ x^2-25 $(2x+6)(2x-6)$ Answer: $4x^2=6x-6x+36$	$(x+5)(x-5)$ $x^2-5x+5x-25$ $\boxed{x^2-25}$

Example	Notes
4 & 5. Multiply.	$(2x^2 + 3y)(2x^2 \cdot 3y)$
$(2x^2 + 3y)(2x^2 - 3y)$	$4x^4 - 6x^2y + 6x^2y - 9y^2$
$4x^4 - 9y$	$\boxed{4x^4 - 9y^2}$
$\left(\dfrac{1}{4}x - \dfrac{2}{3}\right)\left(\dfrac{1}{4}x + \dfrac{2}{3}\right)$	$\left(\dfrac{1}{4}x - \dfrac{2}{3}\right)\left(\dfrac{1}{4}x + \dfrac{2}{3}\right)$
$\dfrac{1}{16}y^2 - \dfrac{4}{9}$	$\boxed{\dfrac{1}{16}x^2 - \dfrac{4}{9}}$

Helpful Hints

Remember that the sum and the difference have to be of the same terms. For example, $(a+b)(a-b) = a^2 - b^2$, but $(a+b)(c-d) \neq ac - bd$.

Remember, when squaring a fraction, you square the numerator and square the denominator.

Concept Check

1. Multiply $(x + y)(x - y)(x^2 + y^2)$ without using the FOIL method.

Practice

Multiply.

2. $(x+4)(x-4)$

3. $(6x+9)(6x-9)$

4. $(8x^2 + 11y)(8x^2 - 11y)$

5. $\left(\dfrac{4}{5}n - \dfrac{2}{7}\right)\left(\dfrac{4}{5}n + \dfrac{2}{7}\right)$

Multiplying Polynomials
Topic 22.5 Squaring Binomials

Vocabulary
a binomial squared   •   binomial   •   FOIL method

1. The rule <u>a binomial squared</u> states that $(a+b)^2 = a^2 + 2ab + b^2$ and

   $(a-b)^2 = a^2 - 2ab + b^2$ where a and b are numbers or algebraic expressions.

Step-by-Step Video Notes
Watch the Step-by-Step Video lesson and complete the examples below.

Example	Notes
1. Simplify $(2x-3)^2$. Write the expression as a multiplication problem, then multiply the binomials. $4x^2 - 12x + 9$ $(2x-3)(2x-3)$ $4x^2 - 6x - 6x - 9$ Answer:	$(2x-3)(2x-3)$ $4x^2 - 6x - 6x + 9$ $\boxed{4x^2 - 12x + 9}$
2. Simplify $(3x+5)^2$. $9x^2 + 30x + 25$ $(3x+5)(3x+5)$ $9x^2 + 15x + 15x + 25$ Answer:	$(3x+5)(3x+5)$ $9x^2 + 15x + 16x + 25$ $\boxed{9x^2 + 30x + 25}$

Example	Notes

4. Simplify $(2x^2 + 3y)^2$. $2y^2 + 3y$

$4x^4 + 6x^2y + 6x^2y + 4y^2$

Answer: $4y^4 + 12x^2y + 4y^2$

$(2x^2 + 3y)(2x^2 + 3y)$

$4x^4 + 6x^2y + 6x^2y + 9y^2$

$\boxed{4x^4 + 12x^2y + 9y^2}$

5. Simplify $\left(\dfrac{1}{4}x - \dfrac{2}{3}\right)^2$.

$\dfrac{1}{16}x^2 - \dfrac{4}{3} + \dfrac{4}{9}$

Answer :

$\left(\dfrac{1}{4}x - \dfrac{2}{3}\right)^2 \quad a - b^2 = a^2 - 2ab +$

$\left(\dfrac{1}{4}x\right)^2 - 2\left(\dfrac{1}{4}x\right)\left(\dfrac{2}{3}\right) + \left(\dfrac{2}{3}\right)^2$

$\dfrac{1}{16}x^2 - \dfrac{4}{12}x + \dfrac{4}{9}$

$\boxed{\dfrac{1}{16}x^2 - \dfrac{1}{3}x + \dfrac{4}{9}}$

Helpful Hints

When you square a binomial, the middle term of the product will always be double the product of the terms of the binomial. The product is called a perfect square trinomial.

The Binomial Squared rule is helpful because it is only necessary to find the squares of the two terms in the binomial and twice their product. Using the FOIL method yields the same result, but requires more calculation and simplification.

Concept Check

1. What will be the middle term when you square the binomial $(3x - 4)$?

Practice

Simplify.

2. $(x + 7)^2$

3. $(5x - 9)^2$

4. $(14x^3 + 13y)^2$

5. $\left(\dfrac{3}{7}y - \dfrac{1}{11}\right)^2$

Name: Sydney

Instructor: Schnabel

Date: 1/25/16

Section: _____

Dividing Polynomials and More Exponent Rules
Topic 23.1 The Quotient Rule

Vocabulary

quotient rule • prime factors method • Zero as an Exponent Property

1. The ___quotient rule___ states that for all non-zero numbers x, $\dfrac{x^a}{x^b} = x^{a-b}$.

2. The ___zero as an Exponent property___ states that for all non-zero numbers x, $\dfrac{x^a}{x^a} = x^0 = 1$.

Step-by-Step Video Notes

Watch the Step-by-Step Video lesson and complete the examples below.

Example	Notes
2. Simplify $\dfrac{25x^6}{10x^3}$. Write the numerator and denominator in an expanded form, to show the factors. Answer:	$\dfrac{5 \cdot 5 \cdot x \cdot x \cdot x \cdot x \cdot x \cdot x}{2 \cdot 5 \cdot x \cdot x \cdot x}$ $\dfrac{5x^3}{2}$
5. Simplify $-\dfrac{16s^6}{32s^2}$. Divide the number parts by the GCF, then use the quotient rule to simplify the variable part. Answer:	$-\dfrac{1}{2}s^4$

Example	Notes
6. Simplify $\dfrac{7^{10}}{7^4}$ using the quotient rule. 7^6	7^{10-4} $=7^6$
Answer:	
9. Evaluate $\dfrac{(ax)^4}{(ax)^4}$.	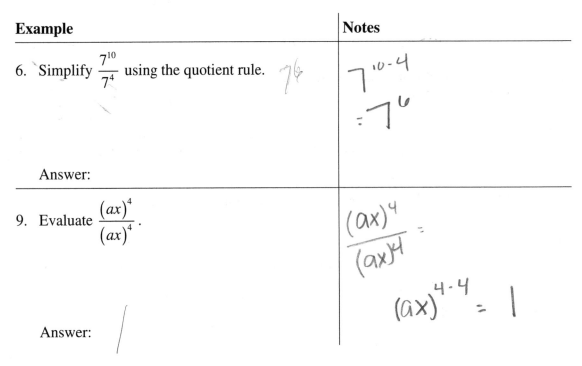 $\dfrac{(ax)^4}{(ax)^4}=$ $(ax)^{4-4}=1$
Answer: /	

Helpful Hints

The quotient rule says to divide like bases, keep the base and subtract the exponents.

Any non-zero number raised to an exponent of zero is equal to 1. 0^0 is undefined.

Concept Check

1. Simplify $\dfrac{a^3}{a^0}$ in two different ways.

Practice

Simplify by the prime factors method.

2. $\dfrac{x^8}{x^4}$

3. $-\dfrac{42x^8}{77x^5}$

Simplify using the quotient rule.

4. $\dfrac{x^{49}}{x^7}$

5. $\dfrac{(37np)^9}{(37np)^9}$

Dividing Polynomials and More Exponent Rules
Topic 23.2 Integer Exponents

Vocabulary

negative exponent • power rule • product rule

1. By definition of a _negative exponent_, $x^{-n} = \dfrac{1}{x^n}$, if $x \neq 0$.

Step-by-Step Video Notes

Watch the Step-by-Step Video lesson and complete the examples below.

Example	Notes
1–3. Simplify. Write your answers with positive exponents. z^{-6} $\dfrac{x^3}{x^7}$ $(x^{-5})(x^3)$	$z^{-4} = \boxed{\dfrac{1}{z^4}}$ $x^{3-7} = x^{-4} = \boxed{\dfrac{1}{x^4}}$ $x^{-5+3} = x^{-2} = \boxed{\dfrac{1}{x^2}}$
4–6. Simplify. Write your answers with positive exponents. 2^{-5} $\dfrac{1}{32}$ -5^{-2} $\dfrac{-1}{25}$ $(-3)^{-3}$ $\dfrac{-1}{21}$	$\overset{4}{\underset{-2\cdot 2\cdot 2\cdot 2}{\wedge}} \overset{4\cdot -2}{\underset{\cdot -2}{\wedge}} = -32 = \boxed{\dfrac{1}{32}}$ $-25 = \boxed{\dfrac{-1}{25}}$ $(-3)^{-3} = \dfrac{1}{(-3)^3} = \dfrac{1}{-27} = \boxed{\dfrac{-1}{27}}$

Example	Notes
8. Simplify $\dfrac{x^{-4}}{y^{-2}}$. Write your answer with positive exponents. Answer:	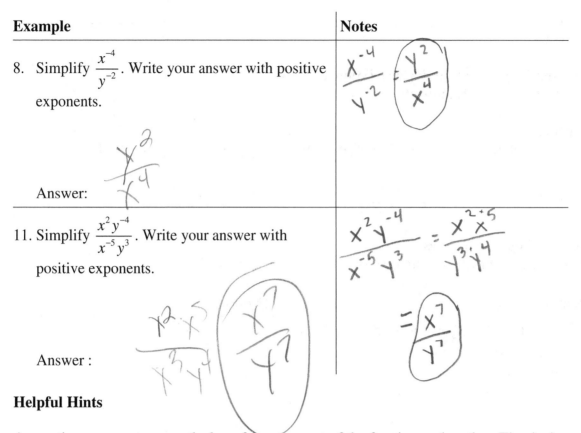
11. Simplify $\dfrac{x^2 y^{-4}}{x^{-5} y^3}$. Write your answer with positive exponents. Answer :	

Helpful Hints

A negative exponent moves the base from one part of the fraction to the other. That is, it moves a numerator to the denominator, and a denominator to the numerator, and the exponent becomes positive.

Concept Check

1. Which is greater, $\left(\dfrac{1}{2}\right)^1$, or $\left(\dfrac{1}{2}\right)^{-1}$?

Practice

Simplify. Write your answers with positive exponents.

2. $2x^{-7}$

3. $\dfrac{x^{-3}}{x^{-5}}$

Evaluate. Write your answers with positive exponents.

4. $\dfrac{(2x)^{-5}}{y^{-10}}$

5. $\dfrac{-3m^{-3}\left(5n^4\right)^{-2}}{m^{-12} n^7}$

Dividing Polynomials and More Exponent Rules
Topic 23.3 Scientific Notation

Vocabulary

Powers of ten   •   scientific notation   •   decimal notation

1. A positive number is written in _Scientific notation_ when it is in the form $a \times 10^n$, where $1 \le a < 10$ and n is an integer.

Step-by-Step Video Notes

Watch the Step-by-Step Video lesson and complete the examples below.

Example	Notes
1. Write $23,400,000$ in scientific notation. Move the decimal point to put one non-zero digit to the left of the decimal point. $2\,3\,4\,0\,0\,0\,0\,0$ The point was moved ⬚5 places. Answer: 2.34×10^7	23,400,000 23400000 $\boxed{2.34 \times 10^7}$
3. Write 2.31×10^6 in decimal notation. The positive power of 10 means move the decimal point ⬚6 places to the _right_ . Answer: 2310000	2.31×10^6 2.310000 $\boxed{2,310,000}$ $2,310,000$
6. Write the number in the following application in decimal notation. The volume of a gold atom is 1.695×10^{-23} cubic centimeters. Write this in decimal notation. $0.00000000000000000001695$ Answer:	1.695×10^{-23} 1.695 $\boxed{.00000000000000000001695}$

Example	**Notes**
7 & 8. Simplify.	5.6×10^{20} $56 = 5.6 \times 10^{1}$
$(8\times10^{16})(7\times10^{4})$	$\boxed{5.6 \times 10^{21}}$
$\sim 56\,0000000000000000000000$	
56×10^{20}	
5.6×10^{21}	
$\dfrac{1.5\times10^{7}}{2.5\times10^{2}}$	$\dfrac{1.5\times10^{7}}{2.5\times10^{2}} = \dfrac{1.5}{2.5} \times \dfrac{10^{7}}{10^{2}}$
6×10^{6}	$\dfrac{1.5}{2.5} \times 10 = \dfrac{15}{25} = .6$
6.0×10^{4}	$.6 = 6 \times 10^{1}$
	$6 \times 10^{5-1}$
	$\boxed{= 6 \times 10^{4}}$

Helpful Hints

Write numbers in scientific notation to make a very big or very small number more compact.

To write a number in decimal form, move the decimal point the correct number of places in the appropriate direction. If the power of 10 is positive, the decimal point moves right. If the power of 10 is negative, the decimal point moves left.

Concept Check

1. How many zeros are at the end of the decimal number equivalent to 3.048×10^{7}?

Practice

Write each number in scientific notation.

2. $24,300,000$

3. $(2.5\times10^{-4})(4\times10^{6})$

Write in decimal notation.

4. 3.04×10^{9}

5. 2.763×10^{-8}

Name: _Sydney_   Schnabel   Date: _1/25/16_

Instructor: _____   Section: _____

Dividing Polynomials and More Exponent Rules
Topic 23.4 Dividing a Polynomial by a Monomial

Vocabulary

trinomial   •   monomial   •   polynomial

1. When dividing a polynomial by a monomial, divide each term of the _polynomial_ by the _monomial._

Step-by-Step Video Notes

Watch the Step-by-Step Video lesson and complete the examples below.

Example	Notes
3. Divide $\dfrac{15x^5 + 10x^4 + 25x^3}{5x^2}$. Divide each term in the polynomial by $5x^2$. $3x^3 + 2x^2 + 5x$ Write the sum of the results. Answer:	$\dfrac{15x^5 + 10x^4 + 25x^3}{5x^2}$ $= \dfrac{15x^5}{15x^2} + \dfrac{10x^4}{15x^2} + \dfrac{25x^3}{15x^2}$ $= \boxed{3x^3 + 2x^2 + 5x}$
4. Divide $\dfrac{63x^7 - 35x^6 - 49x^5}{-7x^3}$. $-9x^4 + 5x^3 + 7x^2$ Answer:	$\dfrac{63x^7 - 35x^6 - 49x^5}{-7x^3}$ $= \dfrac{63x^7}{-7x^3} - \dfrac{35x^6}{-7x^3} - \dfrac{49x^5}{-7x^3}$ $-9x^4 \quad -5x^3 \quad -7x^2$ $= -9x^4 - (-5x^3) - (-7x^2)$ $= \boxed{-9x^4 + 5x^3 + 7x^2}$

Example	Notes
5. Divide $(36x^3 - 18x^2 + 9x) \div (9x)$. $4x^2 - 2x$ Answer:	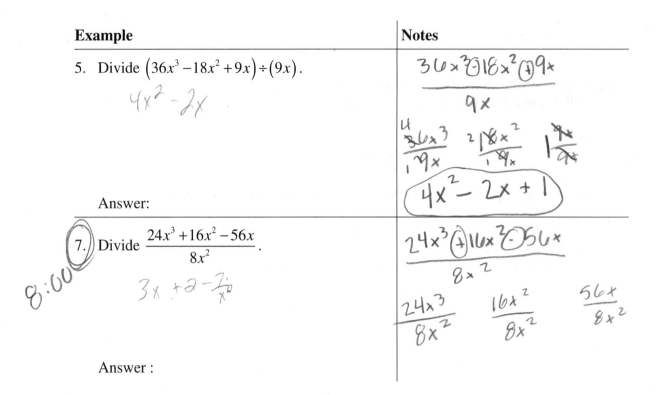
7. Divide $\dfrac{24x^3 + 16x^2 - 56x}{8x^2}$. $3x + 2 - \dfrac{7}{x}$ Answer :	$\dfrac{24x^3 + 16x^2 - 56x}{8x^2}$ $\dfrac{24x^3}{8x^2} \qquad \dfrac{16x^2}{8x^2} \qquad \dfrac{56x}{8x^2}$

$8:00$

Helpful Hints

Split up a fraction into an addition of two or more fractions when trying to simplify the fraction, or with division of a polynomial by a monomial.

If each term does not divide evenly, simplify each individual fraction.

Concept Check

1. Biff states incorrectly that $(24x^3 + 8x^2) \div (8x^2) = 3x$. What was his error?

Practice

Divide.

2. $\dfrac{72x^8 + 45x^6 + 27x^4}{9x^2}$

4. $\dfrac{-42x^9 + 24x^7 + 18x^5}{-6x^5}$

3. $(150x^4 - 220x^3 + 90x^2) \div (10x^2)$

5. $\dfrac{-33x^8 - 24x^5 + 9x^4}{-9x^4}$

Name: _____   Date: _____
Instructor: _____   Section: _____

Dividing Polynomials and More Exponent Rules
Topic 23.5 Dividing a Polynomial by a Binomial

Vocabulary
long division   •   polynomial long division   •   quotient

1. When setting up _____ long division _____, place the terms of the polynomials in descending order. Insert a zero for any missing terms.

Step-by-Step Video Notes
Watch the Step-by-Step Video lesson and complete the examples below.

Example	Notes
1. Divide $(6x^2 + 7x + 2) \div (2x + 1)$. Set the problem up using the long-division symbol. Divide $6x^2$ by $2x$. This is the first term of the answer. $$\begin{array}{r} \boxed{3x} \\ 2x+1\overline{)6x^2 + 7x + 2} \\ 6x^2 + 3x \end{array}$$ Answer: $3x+2$	put renander owr orginal diso
2. Divide $(x^3 + 5x^2 + 11x + 4) \div (x + 2)$. $$\begin{array}{r} \boxed{x} \\ x+2\overline{)x^3 + 5x^2 + 11x + 4} \end{array}$$ Answer: $x^2 + 3x + 5 \frac{-6}{x+2}$	

Example	Notes
3. Divide $(5x^3 - 24x^2 + 9) \div (5x + 1)$.	it missing a factor put 0 and whatever the missing factor is

$5x^3 - 24x^2 + 0x + 9$

Answer: $x^2 - 5x + 1 \dfrac{8}{5x+1}$

Helpful Hints

After setting up a polynomial long division problem, divide the first term of the dividend by the first term of the divisor. The result is the first term of the answer. Then proceed as you would with numbers until the degree of the remainder is less than the degree of the divisor.

Concept Check

1. What missing term must you insert in the dividend when dividing $(p^3 - p + 8) \div (p - 4)$?

Practice

Divide.

2. $(6x^2 + 4x - 10) \div (3x + 5)$ 3. $(28x^2 - 15x - 20) \div (4x + 3)$ 4. $(n^3 - n^2 - 4) \div (n - 2)$

Name: Sydney Schnabel

Date: 1/29/16

Instructor: Schnabel

Section: _____

Factoring Polynomials
Topic 24.1 Greatest Common Factor

Vocabulary

factor • greatest common factor • factoring a polynomial

prime polynomial • distributive property

1. _Factoring a polynomial_ is the process of writing a polynomial as a product of two or more factors.

2. When two or more numbers, variables, or algebraic expressions are multiplied, each is called a _factor_.

Step-by-Step Video Notes

Watch the Step-by-Step Video lesson and complete the examples below.

Example	Notes
1–3. Find the GCF.	① $GCF = x^3$
$x^3, x^7,$ and x^5 $x^3 \qquad x^3(x^4 \text{ and } x^2)$	
	② $GCF = y$
$y, y^4,$ and y^7 $y(y^3 \text{ and } y^6)$	
x and y^2	③ $GCF = 1$
5. Factor out the GCF of $9x^5 + 18x^2 + 3x$. Write each term as the product of the GCF and each term's remaining factors. $3x(3x^4 + 6x + 1)$	$\begin{matrix} 9 & 18 & 3 \\ 3\ 3 & 3\ 6 & \\ & 2\ 3 & \end{matrix}$ $GCF = 3x$ $9x^5 + 18x^2 + 3x$ $\boxed{= 3x(3x^4 + 6x + 1)}$
Answer:	

Example	Notes
6. Factor out the GCF of $8x^3y + 16x^2y^2 - 24x^3y^3$.	$8x^2y = GCF$
$8x^3y(\cancel{2}x - 4xy^2)$	$8x^2y(x + 2y - 3xy^2)$
Answer:	

7 & 8. Factor out the GCF.

$24ab + 12a^2 + 36a^3$	⑦ GCF: $12a$
$12a(2b + a + 3a^2)$	$12a(2b + a + 3a^2)$
$3x + 7y + 12xy$	⑧ GCF = 1
$1(3x + 7y + 12xy)$	$1(3x + 7y + 12xy)$

Helpful Hints
Factoring a polynomial changes a sum and/or difference of terms into a product of factors.

To find the GCF of two or more terms with coefficients and variables, find the product of the GCF of the coefficients and the GCF of the variable factors.

Concept Check
1. Allison states incorrectly that $3x + 24y - 15z$ is a prime polynomial because the variables in the terms are different. What is her error?

Practice
Factor out the GCF.

2. $72x^8 - 54x^6 + 27x^5$

4. $-42x^9y + 24x^7y^2 + 18x^5y^3$

3. $140x^4 - 210x^3 + 70x^2$

5. $-33x^8y^8 - 24x^5y^5 + 9x^4y^3$

Name: Sydney

Instructor: Schnabel

Date: 1/29/16

Section: _____

Factoring Polynomials
Topic 24.2 Factoring by Grouping

Vocabulary
common factor • factoring by grouping • polynomial

1. When __factoring by grouping__, collect the terms into two groups so that each group has a common factor. Then factor out the GCF from each group so that the remaining factor in each group is the same.

Step-by-Step Video Notes
Watch the Step-by-Step Video lesson and complete the examples below.

Example	Notes
1. Factor $2x^2 + 3x + 6x + 9$ by grouping.	$(2x^2 + 3x)(+6x + 9)$
Group terms. $x(2x+3) + 3(2x+3)$	$x(2x+3) \ 3(2x+3)$
Factor within groups.	$(x+3)(2x+3)$
Factor the entire polynomial.	Check: $2x^2 + 3x + 6x + 9$
Multiply to check your answer.	
Answer:	
2. Factor $2x^2 + 5x - 4x - 10$ by grouping.	$(2x^2 + 5x)(-4x-10)$
$x(2x+5) - 2(2x+5)$	$x(2x+5) -2(2x+5)$
	$(x-2)(2x+5)$
	$2x^2 + 5x - 4x - 10$
Answer:	

Example	Notes
3. Factor $2ax - a - 2bx + b$ by grouping. $2a(x-1) - 2b(x-1)$ Answer:	$(2ax - a)(-2bx + b)$ $a(2x+1) - b(2x-1)$ $(a - b(2x-1)$ $2ax - a - 2xb + b$
4. Factor $10x^2 - 8xy + 15x - 12y$ by grouping. $2(5x-4y)\ 3(5x-4y)$ Answer:	$(10x^2 - 8xy) + (15x - 12y)$ $2x(5x-4y)\quad 3(5x-4y)$ $(2x+3)(5x-4y)$ $10x^2 - 8xy + 15x - 12y$

Helpful Hints

Sometimes you will need to factor out a negative common factor from the second two terms to obtain two terms that contain the same binomial factor. When factoring out a negative, check your signs carefully.

Concept Check

1. Can you factor Example 1, $2x^2 + 3x + 6x + 9$, by grouping the terms differently? If so, do you get the same factors?

Practice

Factor by grouping.

2. $6x^2 - 8x + 9x - 12$

3. $6x + 18y + ax + 3ay$

4. $4x + 20y - 3ax - 15ay$

5. $16a^2 - 14ab + 24a - 21b$

Factoring Polynomials

Topic 24.3 Factoring Trinomials of the Form $x^2 + bx + c$

Vocabulary

trinomial   •   FOIL method   •   prime polynomial

1. A ___prime polynomial___ is a polynomial that cannot be factored.

Step-by-Step Video Notes

Watch the Step-by-Step Video lesson and complete the examples below.

Example	Notes

1. Factor $x^2 + 7x + 12$.
 	
Write the first two terms of the binomial factors.
$(x+3)(x+4)$
 	
List the possible pairs of factors of 12.
 	
Answer: $(x+3)(x+4)$
	$(x+3)(x+4)$ $12: 1,12$
	                         $2,6$
	                         $3,4$
	
	$12: 1,12; 2,6; 3,4$
	$-1,-12;-2,6;-3,4$

2. Factor $x^2 - 8x + 15$. Write the first two terms of the binomial factors. List the possible pairs of factors of 15. $(x-5)(x-3)$ Answer: $(x-3)(x-5)$	$15: 1,15, 3,5$ $-1,-15,-3,-5$ $(x+^{-3})(x+^{-5}) = (x-3)(x-5)$ $(x-3)(x-5)$ $x^2 - 5x - 3x + 15$ $x^2 - 8x + 15$

287

Example	Notes
3. Factor $x^2 - 3x - 10$. $(x-5)(x+2)$ Answer:	$10: -1, 10 , -2,5$ $1, -10 , 2, -5$ $(x+2)(x + -5)$ $(x+2)(x-5)$
5. Factor $4x^2 + 40x - 96$. First factor out the GCF. $2(x^2 + 12x + 20)$ $2(x+2)(x+10)$ Answer:	$4(x^2 + 10x - 24)$ $4(x + -2)(x + 12)$ $4(x-2)(x+12)$ $24 = -1, 24 ; -2, 12 ; -3 ; 8$ $1, -24 ; 2, -12 ; 3 -8$ $-4, 6$ $4 ; -6$

Helpful Hints

Trinomials of the form $x^2 + bx + c$ factor as $(x + \square)(x + \square)$. Trinomials of the form $x^2 - bx + c$ factor as $(x - \square)(x - \square)$, and trinomials of the form $x^2 + bx - c$ factor as $(x + \square)(x - \square)$.

Concept Check

1. To factor a trinomial of the form $x^2 + bx + c$ as the product of two binomials, what must be the product and the sum of the constant terms of the binomial factors?

Practice

Factor.

2. $x^2 + 9x + 18$

4. $3x^2 + 39x + 120$

3. $x^2 - 10x + 24$

5. $x^2 + 13x - 48$

Factoring Polynomials
Topic 24.4 Factoring Trinomials of the Form $ax^2 + bx + c$

AC Method

Vocabulary
FOIL method   •   reverse FOIL method

1. To factor the trinomial $ax^2 + bx + c$ using the _____, list the different factorizations of ax^2 and c. List the possible factoring combinations until the correct middle term is reached.

Step-by-Step Video Notes
Watch the Step-by-Step Video lesson and complete the examples below.

Example	Notes
1. Factor $2x^2 + 5x + 3$. *6x* List the different factorizations of $2x^2$ and 3. $2x^2 + 3x + 2x + 3$ $x(2x+3)$ $1(2x+3)$ List the possible factoring combinations and check the middle term of each combination. Answer: $(x+1)(2x+3)$	
2. Factor $4x^2 - 21x + 5$. List the different factorizations of $4x^2$ and 5. $4x^2 - 20x - 1x + 5$ $4x(x-5) - 1(x-5)$ Answer: $(4x-1)(x-5)$	

Example	Notes
3. Factor $10x^2 - 9x - 9$.	

$(10x^2 - 15x) + (6x - 9)$

$5x(2x-3)\ 3(2x-3)$

Answer: $(5x+3)(2x-3)$

4. Factor $3x^2 + 5x - 12$.	

$3x^2 + 9x)(4x - 12$

$3x(x+3)\,4(x+3)$

Answer: $(3x-4)(x+3)$

Helpful Hints

When factoring trinomials of the form $ax^2 + bx + c$, there can be many combinations of the factors of ax^2 and c. If the trinomial can be factored, only one of these combinations gives the correct middle term.

If c is positive, the signs of both of its factors are the same as the sign of b. If c is negative, then the signs of both of its factors are different.

Concept Check

1. Does the trinomial $8x^2 + 26x + 15$ factor to $(4x+5)(2x+3)$? How can you tell without multiplying the binomials completely?

Practice

Factor.

2. $3x^2 + 7x + 2$ 4. $9x^2 + 26x + 16$

3. $10x^2 - 37x + 7$ 5. $-x^2 - 5x + 24$

Factoring Polynomials
Topic 24.5 More Factoring of Trinomials

Vocabulary
greatest common factor • reverse FOIL method • prime polynomial

1. When factoring a trinomial of the form $ax^2 + bx + c$, if c does not have any factors that have a sum of b, then the trinomial cannot be factored and is called a

_____.

Step-by-Step Video Notes
Watch the Step-by-Step Video lesson and complete the examples below.

Example	Notes
1. Factor $9x^2 + 3x - 30$. Factor out the GCF, then factor the trinomial. $3(3x^2 + x - 10)$ $3(3x^2 + 6x) - 5x - 10$ $3x(x+2) - 5(x+2)$ Answer: $3(3x-5)(x+2)$	
2. Factor $3 - 10x + 8x^2$. $8x^2 - 10x + 3$ $8x^2 - 6x - 4x + 3$ $2x(4x-3) - 1(4x-3)$ Answer: $(2x-1)(4x-3)$	

Example	**Notes**
3. Factor $5x^2 + 7x + 4$. *(handwritten: 20, 1 02, 5 4)* Answer: *prime*	
4. Factor $2x^2 + 12x + 24$. *(handwritten: $2(x^2 + 6x + 12)$)* Answer: $2\left(x^2 + 6x + 12\right)$	

Helpful Hints

When factoring trinomials of the form $ax^2 + bx + c$, always look first for a greatest common factor (GCF). This may be the only possible factoring that can be done. Do not forget to include the GCF in your final answer.

Concept Check

1. Find a whole number value of b between 10 and 20 that would make the trinomial $3x^2 + bx + 16$ a prime polynomial.

Practice

Factor.

2. $8x^2 + 8x - 30$

4. $42x + 20x^2 + 2x^3$

3. $42x^3 - 45x^2 + 12x$

5. $7x^3 + 21x^2 - 14x$

Factoring and Quadratic Equations
Topic 25.1 Special Cases of Factoring

Vocabulary
difference of two squares   •   perfect square number   •   binomial squared
perfect square trinomial

1. In a _perfect square number_, the first and last terms are perfect squares, and the
   middle term is twice the products of square roots of the first and last terms.

Step-by-Step Video Notes
Watch the Step-by-Step Video lesson and complete the examples below.

Example	Notes
2–4. Determine if the expression is a difference of two squares.	2. $x^2 - 16$ $(x)^2 - (4)^2$ yes
$x^2 - 16$ yes $(x)^2-(4)^2$	
$x^2 - 7$ No	3. No
$4x^2 + 81$ no	4. $4x^2 + 81$ no
5 & 6. Factor. Remember the property $a^2 - b^2 = (a+b)(a-b)$.	5. $x^2 - 49$ $(x)^2 - (7)^2 = (x+7)(x-7)$
$x^2 - 49$ $x+7$ $x-7$	
$25b^2 - 64$ $(5b)^2 - (8)^2$ $5b + 8$ $5b - 8$	6. $25b^2 - 64$ $(5b)^2 - (8)^2$ $(5b+8)(5b-8)$

Example	Notes
7 & 8. Factor.	7. $4x^2 - 81y^2$
	$(2x)^2 - (9y)^2$
$4x^2 - 81y^2$ $(2x - 9y)(2x + 9y)$	$(2x + 9y)(2x - 9y)$
$-9x^2 + 1$ $(3x + 1)(-3x + 1)$ $(1 + 3x)(1 - 3x)$	8. $-9x^2 + 1$ $(1)^2 - (3x)^2$
	$1 - 9x^2$ $(1 + 3x)(1 - 3x)$

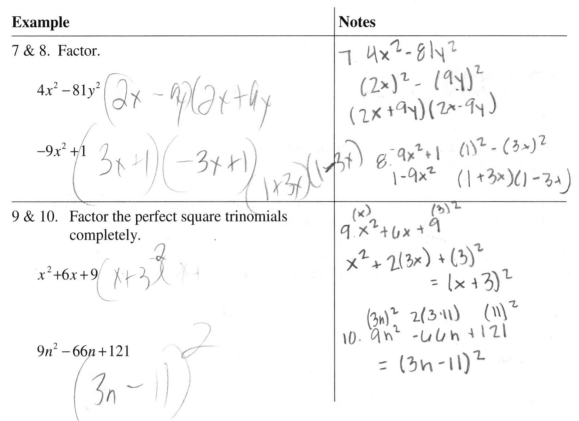

9 & 10. Factor the perfect square trinomials completely.

$x^2 + 6x + 9$ $(x + 3)^2$

9. $x^2 + 6x + 9$ $\overset{(x)}{x^2} + 6x + \overset{(3)^2}{9}$

$x^2 + 2(3x) + (3)^2$
$= (x + 3)^2$

$9n^2 - 66n + 121$ $(3n - 11)^2$

10. $\overset{(3n)^2}{9n^2} \overset{2(3 \cdot 11)}{- 66n} + \overset{(11)^2}{121}$
$= (3n - 11)^2$

Helpful Hints
Other than possibly having a GCF, a sum of two squares will not factor.

If the middle term in a perfect square trinomial is being subtracted, the sign between the terms of the binomial factors will be a minus sign.

Concept Check
1. Can $x^2 - \dfrac{1}{4}$ be factored as a difference of squares? Explain.

Practice
Factor completely.

2. $x^2 - 144$

4. $16m^2 - 40m + 25$

3. $25x^2 - 81y^2$

5. $3x^2 - 42x + 147$

Factoring and Quadratic Equations
Topic 25.2 Factoring Polynomials

Vocabulary

a difference of two squares   •   perfect square trinomial
greatest common factor   •   reverse FOIL method

1. When factoring a polynomial, always start by looking for a _greatest common factor_.

Step-by-Step Video Notes
Watch the Step-by-Step Video lesson and complete the examples below.

Example	Notes
1. Factor $3k^2 - 48$ completely. Factor out the GCF, if possible. $3(\text{ }) - 3(16\text{ })$ $3(k^2 - 16)$ Factor the remaining binomial factor. $3(k^2 - 16)$ $3(k+4)(4-4)$ Answer:	$3k^2 - 48$ $3(k^2 - 16)$ $3(k+4)(k-4)$ $\boxed{3(k+4)(k-4)}$
2. Factor $12n^3 - 12n^2 - 144n$ completely. $12n(n^2 - n - 12)$ $12n(n+3)(nf-4)$	$12n^3 - 12n^2 - 144n$ $12n(n^2 - n - 12)$ $12n(n+3)(n-4)$ $\boxed{12n(n+3)(n-4)}$

Example	Notes
4. Factor $x^3 + 2x^2 - 9x - 18$ completely. 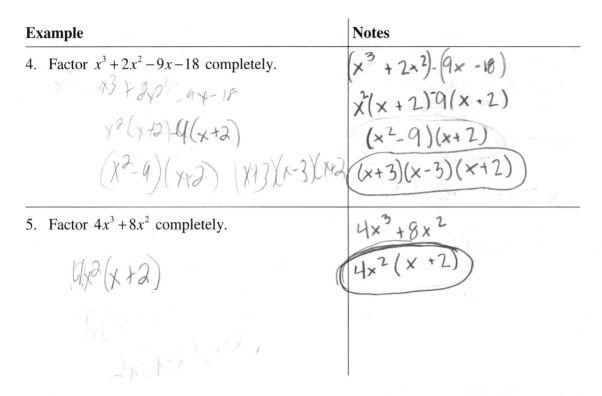 $x^3 + 2x^2 - 9x - 18$ $x^2(x+2) - 9(x+2)$ $(x^2-9)(x+2)$ $(x^2-9)(x+2)$ $(x+3)(x-3)(x+2)$	$(x^3 + 2x^2) \cdot (9x - 18)$ $x^2(x+2)^-9(x+2)$ $(x^2-9)(x+2)$ $\boxed{(x+3)(x-3)(x+2)}$
5. Factor $4x^3 + 8x^2$ completely. $4x^2(x+2)$	$4x^3 + 8x^2$ $\boxed{4x^2(x+2)}$

Helpful Hints

When factoring polynomials, check for special cases and use different strategies depending on how many terms are in the polynomial. The polynomial will be factored completely when each factor is a prime polynomial.

Concept Check

1. If a polynomial has four terms and no GCF, how should you try to factor?

Practice

Factor completely.

2. $9x^6 - 48x^3 + 64$ 4. $64n^8 - 4$

3. $-3x^3 + 18x^2 + 48x - 288$ 5. $15x^2 - 23x + 4$

Factoring and Quadratic Equations
Topic 25.3 Solving Quadratic Equations by Factoring

Vocabulary
Zero Property of Multiplication   •   quadratic equation   •   standard form

1. A _quadratic equation_ is an equation of the form $ax^2 + bx + c = 0$ where $a \neq 0$.

Step-by-Step Video Notes
Watch the Step-by-Step Video lesson and complete the examples below.

Example	Notes
1. Solve $x^2 + 4x = 0$ by factoring. $ab=0$ Factor completely. $x(x+4)=0$ $x=0$ $x+4=0$ Set each factor equal to zero. $x=0$ $x=-4$ Solve each equation. Answer: $x=0$ $x=-4$	$x(x+4)=0$ $\boxed{x=0}$ $x+4=0$ $\boxed{x=-4}$
3. Solve $10x^2 - x = 2$ by factoring. $10x^2 \; x=2$ $10x^2 \; x-2=0$ $(5x+2)(2x-1)$ Answer: $-\frac{2}{5}, \frac{1}{2}$	always set to $x=0$ $10x^2 - x = 2$ $10x^2 - x - 2 = 0$ $\diagup\!\!\!\!\diagdown \begin{smallmatrix}-20\\-1\end{smallmatrix}$ $(x-5)(x+4)$ $(10x^2 - 5x) + (4x - 2)$ $5x(2x-1)\,2(2x-1)$ $2x-1=0$ $(5x+2)(2x-1)=0$ $2x=1$ $5x+2=0$ $\boxed{x=\frac{1}{2}}$ $5x=-2$ $\boxed{x=-\frac{2}{5}}$

Example	Notes
5. Solve $4x^2 + 9 = 12x$ by factoring. $$4y^2 - 12x + 9 = 0$$ $$(2y-3)^2$$ $$(2x-3)(2x-3) = 0$$ $$x = \frac{3}{2}$$	$$4x^2 + 9 - 12x = 0$$ $$4x^2 - 12x + 9 = 0 \qquad \frac{36}{-6 \; -6} \; \frac{}{12}$$ $$(4x^2 - 6x) - (6x + 9) = 0 \qquad 2x - 3 = 0$$ $$2x(2x-3) \; 3(2x-3) \qquad 2x = 3$$ $$(2x+3)(2x-3) \qquad \boxed{x = \frac{3}{2}}$$ $$2x+3=0 \quad 2x=-3$$
6. Solve $x^2 - 64 = 0$ by factoring. $$(x+8)(x-8)$$ $$x+8=0 \qquad x-8=0$$ $$x=8 \qquad x=8$$ $$x = 8, -8$$	$$x^2 - 64 = 0$$ $$(x+8)(x-8) = 0$$ $$x+8=0 \qquad x-8=0$$ $$\boxed{x=-8} \qquad \boxed{x=8}$$

Helpful Hints

The highest degree of any term in a quadratic equation is 2.

The solutions to quadratic equations are also called roots or zeros. A quadratic equation has at most two solutions. Sometimes both solutions will be the same number.

Concept Check

1. Write the quadratic equation $25x^2 + 34x = 4x - 9$ in standard form.

Practice
Solve by factoring.

2. $7x^2 - 28x = 0$

3. $3x^2 - 3x - 126 = 0$

4. $24x^2 + 2x = 35$

5. $25x^2 = 80x - 64$

Name: Sydney
Instructor: Schnabel

Date: 2/4/16
Section: _____

Factoring and Quadratic Equations
Topic 25.4 Applications

Vocabulary
quadratic equation • roots of a quadratic equation

1. When using a _____ to solve for real-world measurements such as time, distance, length, etc., only use the positive roots of the equation

Step-by-Step Video Notes
Watch the Step-by-Step Video lesson and complete the examples below.

Example	Notes
1. The cliff diver jumps from a platform placed on a cliff approximately <u>144 feet above the surface</u> of the sea. Disregarding air resistance, the height S, in feet, of a cliff diver above the ocean after t seconds is given by the quadratic equation $S = -16t^2 + 144$. How long does it take the diver to reach the water? (Note: The height when he hits the water is 0 feet.) Answer: ③	$0 = -16t^2 + 144$ $16t^2 - 144 = 0$ $16(t^2 - 9) = 0$ $16(t+3)(t-3) = 0$ $t+3 = 0 \quad t-3 = 0$ $\cancel{-3} \quad \boxed{t = 3}$
2. A tennis ball is thrown upward with an initial velocity of 8 meters/second. Suppose that the initial height above the ground is 4 meters. Find the height S of the ball after 1 second. At what time t will the ball hit the ground? Remember, the equation is $S = -5t^2 + vt + h$. $0 = -5t^2 + 8t + 4$ Answer: the height at $S=7$ $t=2$	$S = -5t^2 + 8t + 4$ $= -5(1)^2 + 8(1) + 4$ $= -5 + 8 + 4$ $S = 7$ $t=1; S=7$ $5t^2 - 8t - 4 = 0$ $(5t^2 - 10t)(2t - 4)$ $5t(t-2)2(t-2)$ $(5t+2)(t-2)$ $5t+2=0 \quad t-2=0$ $5t=-2 \quad \boxed{t=2}$

299
Copyright © 2012 Pearson Education, Inc.

Example	Notes
4. The length of the base of a rectangle is 7 inches greater than the height. If the total area of the rectangle is 120 square inches, what are the length of the base and height of the rectangle? For a rectangle, Area = base · height.	$120 = (h+7)(h)$ $120 = h^2 + 7h$ $0 = h^2 + 7h - 120$ $(h^2 + 15h) - (8h - 120)$ $h(h+15) - 8(h+15)$ $(h-8) = 0 \quad h + 15 = 0$ $\boxed{h = 8} \quad \cancel{h = -15}$

height = 8
bas = 15
$8 + 7 = 15$

Answer: *8 in*
15 in

Helpful Hints

The process of solving applications involving quadratic equations is the same as solving quadratic equations in general, with the exception that sometimes there are application specific questions that need to be answered based on the solutions.

Concept Check

1. Write the quadratic equation $3200 = -16t^2 + 480t$ in standard form with a positive coefficient for t^2.

Practice

Solve.

2. An egg is thrown upward with an initial velocity (v) of 9 meters/second. Suppose that the initial height (h) above the ground is 2 meters. At what time t will the egg hit the ground? Use the quadratic equation $S = -5t^2 + vt + h$.

3. A rocket is fired upwards with a velocity (v) of 640 feet per second. Find how many seconds it takes for the rocket to reach a height of 6,400 feet. Use the quadratic equation $S = -16t^2 + 640t$.

4. The length of the base of a rectangle is 4 inches more than twice the height. If the total area of the rectangle is 126 square inches, what are the lengths of the base and height of the rectangle? Remember that for a rectangle, Area = base · height.

Introduction to Rational Expressions
Topic 26.1 Undefined Rational Expressions

Vocabulary

rational number • rational expression • zero property of multiplication

1. Any ___rational number___ can be written as a fraction of two algebraic expressions.

2. Any ___rational expression___ can be written as a fraction of two integers.

Step-by-Step Video Notes
Watch the Step-by-Step Video lesson and complete the examples below.

Example	Notes
1. Find any values of the variable that will make the rational expression $\dfrac{15x^2 + 25x}{5x}$ undefined. Set the denominator equal to zero. Solve. $\dfrac{15(0)^2 + 25(0)}{5(0)}$ $\dfrac{0}{0}$ Answer: undefined for $x=0$	$\dfrac{15x^2 + 25x}{\dfrac{5x}{5} = \dfrac{0}{5}}$ $x = 0$ *undefined for $x = 0$
2. Find the value of the variable that will make the rational expression $\dfrac{24x^2 + 9x}{8x+3}$ undefined. $8x+3=0$ $-3\ -3$ $\dfrac{8x}{8} = \dfrac{3}{8}$ $x = 3$ Answer: undefined for $x = -\dfrac{3}{8}$	$\dfrac{24x^2 + 9x}{8x+3 = 0}$ $8x = -3$ $x = -\dfrac{3}{8}$ *undefined for $x = -\dfrac{3}{8}$

Example	Notes
4. Find any values of the variable that will make the rational expression $\dfrac{x^2+6x+8}{x^2-16}$ undefined.	$\dfrac{x^2+6x+8}{x^2-16}=0$ $(x-4)(x+4)=0$ $x=4,-4$

(handwritten, left:) $x^2+5x+6=0$ $(x+3)(x+2)$ $x=-3+=-2$

(handwritten, center:) $x+4 \; x-4$

Answer: $4, -4$

| 5. Find any values that will make the rational expression $\dfrac{x^3+2x^2+x+2}{x^2+1}$ undefined. | $\dfrac{x^2+2x^2+x+2}{x^2+1}=0$ prime/does not factor |

Answer: *defined for all real numbers*

Helpful Hints

Division by zero is undefined, so the denominator of a rational expression cannot be zero. Any value of the variable that would make the denominator zero is not allowed. Thus, the domain of a rational expression is all values that do not give a zero in the denominator.

Concept Check

1. Is the expression $\dfrac{2x}{x}$ defined for all values of x? Explain.

Practice

Find any values of the variable that will make the rational expression undefined.

2. $\dfrac{68x^3+52x^2}{17x}$

3. $\dfrac{49x^3+56x}{5x-8}$

4. $\dfrac{x^2-8}{x^2-12x+32}$

5. $\dfrac{x^4-25}{x^2+9}$

Introduction to Rational Expressions
Topic 26.2 Simplifying Rational Expressions

Vocabulary

Basic Rule of Fractions • simplifying rational expressions • common factors

1. The ___Basic rule of fractions___ states that for any rational expression $\dfrac{ac}{bc}$ and any

    expressions a, b, and c, (where $b \neq 0$ and $c \neq 0$), $\dfrac{ac}{bc} = \dfrac{a}{b}$.

Step-by-Step Video Notes

Watch the Step-by-Step Video lesson and complete the examples below.

Example	Notes
1. Simplify $\dfrac{21}{39}$.	$\dfrac{21}{39} = \dfrac{3 \cdot 7}{3 \cdot 13} = \boxed{\dfrac{7}{13}}$
Answer: $\dfrac{7}{13}$	
2. Simplify $\dfrac{2x+6}{3x+9}$. Factor the numerator and denominator completely, and divide by common factors.	$\dfrac{2x+6}{3x+9} = \dfrac{2(x+3)}{3(x+3)} = \boxed{\dfrac{2}{3}}$
Answer : $\dfrac{2}{3}$	

Example	Notes
3. Simplify $\dfrac{x^2+9x+14}{x^2-4}$. $\dfrac{(x+7)(x+2)}{(x+2)(x-2)}$ Answer : $\dfrac{x+7}{x-2}$	$\dfrac{x^2+9x+14}{x^2-4}$ $\dfrac{(x+7)(x+2)}{(x+2)(x-2)}=\boxed{\dfrac{x+7}{x-2}}$
5. Simplify $\dfrac{5x^2-45}{45-15x}$. $\dfrac{5(x^2-9)}{3\cancel{15}(3-x)\cancel{-1}}=\dfrac{15\cancel{(x+3)(x-3)}=-1}{}$ Answer : $\dfrac{-1(x+3)}{3}=\dfrac{-1(x+3)}{3\cdot 1}$	$\dfrac{5x^2-45}{45-15x}=\dfrac{5(x^2-9)}{15(3-x)}$ $\dfrac{5(x-3)(x+3)}{15(3-x)}=\dfrac{5\cdot 1}{5\cdot 3}$ $\dfrac{1(x+3)\cdot 1}{3(1)}=\boxed{\dfrac{-1(x+3)}{3}}$

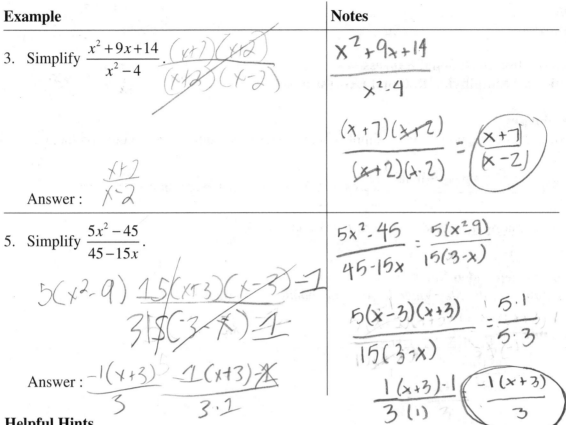

Helpful Hints

Factor the numerator and denominator completely, being aware of special factoring cases like differences of two squares, trinomial squares, monomial factors, and negative factors.

For all polynomials A and B, where $A \neq B$, it is true that $\dfrac{A-B}{B-A}=-1$.

Concept Check

1. What makes the expression $\dfrac{125x^3-9y^2}{9y^2-125x^3}$ easy to simplify? What is the simplified form?

Practice
Simplify.

2. $\dfrac{9x+27}{4x+12}$

4. $\dfrac{x^2+9xy+18y^2}{5x^2+17xy+6y^2}$

3. $\dfrac{5x-4}{12-15x}$

5. $\dfrac{21-4x-x^2}{4x^2-36}$

Introduction to Rational Expressions
Topic 26.3 Multiplying Rational Expressions

Vocabulary

multiplying fractions • multiplying rational expressions • the quotient rule

1. When multiplying rational expressions, use **the quotient rule** to simplify the variable parts.

Step-by-Step Video Notes
Watch the Step-by-Step Video lesson and complete the examples below.

Example	Notes
1. Multiply $\dfrac{12}{7} \cdot \dfrac{49}{36}$ $\dfrac{7}{3}$	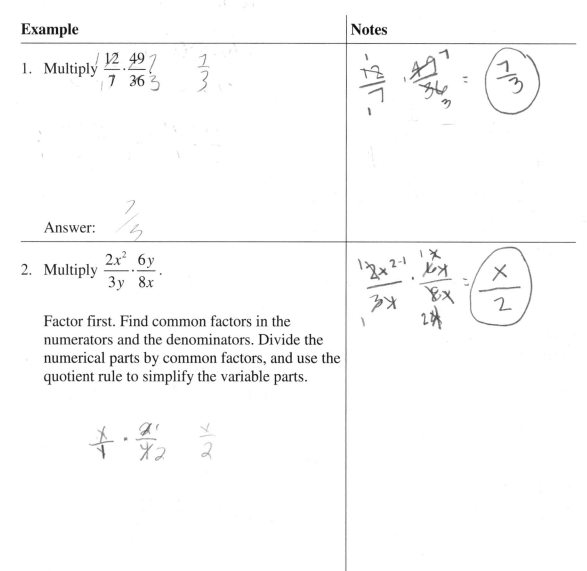
Answer: $\dfrac{7}{3}$	
2. Multiply $\dfrac{2x^2}{3y} \cdot \dfrac{6y}{8x}$.	
Factor first. Find common factors in the numerators and the denominators. Divide the numerical parts by common factors, and use the quotient rule to simplify the variable parts. $\dfrac{x}{1} \cdot \dfrac{2}{2} \quad \dfrac{x}{2}$	
Answer:	

Example	Notes

Example

3. Multiply $\dfrac{7x}{22} \cdot \dfrac{11x+33}{7x+21}$.

(handwritten) $\dfrac{x(x+3)}{2(x+3)}$

Answer:

Notes

$\dfrac{7x}{22} \cdot \dfrac{11x+33}{7x+21}$

$\dfrac{7x}{22} \cdot \dfrac{11(x+3)}{7(x+3)} = \dfrac{7x}{22} \cdot \dfrac{11}{7}$

$= \dfrac{1 \cdot x \cdot 1}{2 \cdot 1 \cdot 1} = \dfrac{x}{2}$

---

5. Multiply $\dfrac{x^2-x-12}{16-x^2} \cdot \dfrac{2x^2+7x-4}{x^2-4x-21}$.

(handwritten) $\dfrac{(x-4)(x+3)}{(4-x)(4+x)} \cdot \dfrac{(2x-1)(x+4)}{(x-7)(x+3)} =$

Answer:

(handwritten Notes column)

$\dfrac{x^2-x-12}{16-x^2} \cdot \dfrac{2x^2+7x-4}{x^2-4x-21}$

$\dfrac{(x-4)(x+3)}{(4-x)(4+x)} \cdot \dfrac{2x(x+4)-1(x+4)}{(x-7)(x+3)}$

$\dfrac{-1 \cdot 1 \cdot (2x-1) \cdot 1}{1 \cdot 1 \cdot (x-7) \cdot 1} = \dfrac{-1(2x-1)}{(x-7)}$

Helpful Hints

To multiply two rational expressions, find the common factors in the numerators and the denominators. Divide the numerators and denominators by common factors. Then multiply the remaining factors.

Concept Check

1. Are there any common factors to divide in the multiplication $\dfrac{4}{x} \cdot \dfrac{x+4}{4x^2+x}$?

Practice
Multiply.

2. $\dfrac{49x^2}{42y^3} \cdot \dfrac{48y^6}{35x}$

4. $\dfrac{x^4-16}{8x^2+32} \cdot \dfrac{32x^2+24x}{3x^3-4x^2-4x}$

3. $\dfrac{3x-24}{6x+75} \cdot \dfrac{4x+50}{12x-96}$

5. $\dfrac{7-x}{x^2-4x-21} \cdot \dfrac{10-x-x^2}{9x-18}$

Introduction to Rational Expressions
Topic 26.4 Dividing Rational Expressions

Vocabulary

dividing fractions   •   dividing rational expressions   •   reciprocal

1. When _diving rational expressions_ multiply the first rational expression by the reciprocal of the second rational expression.

Step-by-Step Video Notes

Watch the Step-by-Step Video lesson and complete the examples below.

Example	Notes
2. Divide $\dfrac{-12x^2}{5y} \div \dfrac{18x}{15y}$. Find the reciprocal of the second rational expression and multiply. Divide the numerators and denominators by common factors and then write the remaining factors as one fraction. Answer: $-2x$	$\dfrac{-12x^2}{5y} \cdot \dfrac{15y}{18x} = \dfrac{-2x}{1}$ $= -2x$
3. Divide $\dfrac{8x}{14} \div \dfrac{8x-32}{7x-28}$. Answer: $\dfrac{1}{2}$	$\dfrac{8x}{14} \cdot \dfrac{7(x-4)}{8(x-4)} = \dfrac{1 \cdot x \cdot 1 \cdot 1}{2 \cdot 1 \cdot 1}$ $= \dfrac{x}{2}$

Example

4. Divide $\dfrac{x^2+3x-10}{x^2+x-20} \div \dfrac{x^2+4x+3}{x^2-3x-4}$.

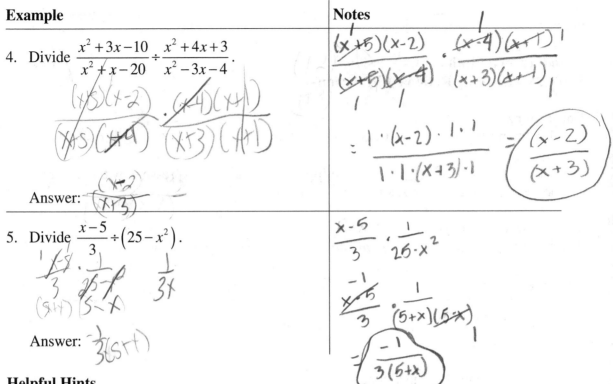

Answer:

5. Divide $\dfrac{x-5}{3} \div \left(25-x^2\right)$.

Answer:

Notes

Helpful Hints

The reciprocal of an integer or an expression is 1 over the integer or the expression. To get the reciprocal of a fraction or a rational expression, invert the fraction or expression.

Remember when multiplying or factoring that $\dfrac{(a-b)}{(b-a)}=-1$.

Concept Check

1. Dividing by $\dfrac{4x}{3+y}$ is the same as multiplying by what rational expression?

Practice

Divide.

2. $\dfrac{8x^4}{45y^3} \div \dfrac{32x^2}{9y^2}$

4. $\dfrac{14x^2+13x+3}{28x^2+5x-3} \div \dfrac{6x^2-7x-5}{12x^2+17x-5}$

3. $\dfrac{11x}{42} \div \dfrac{11x-77}{6x-42}$

5. $\dfrac{3x-5}{6} \div \left(25-9x^2\right)$

Adding and Subtracting Rational Expressions
Topic 27.1 Adding Like Rational Expressions

Vocabulary

like rational expressions   •   unlike rational expressions   •   numerator

1. Rational expressions with a common denominator are called ___like rational expressions___.

Step-by-Step Video Notes
Watch the Step-by-Step Video lesson and complete the examples below.

Example	Notes
1. Add $\dfrac{5a}{4a+2b}+\dfrac{6a}{4a+2b}$. Add the numerators and keep the denominator the same. $\dfrac{11a}{4a+2b}$ Answer:	$\dfrac{5a}{4a+2b}+\dfrac{6a}{4a+2b}=\dfrac{11a}{4a+2b}$
2. Add $\dfrac{-7m}{2n}+\dfrac{m}{2n}$. $\dfrac{-3m}{n}$ Answer: $\dfrac{-3m}{n}$	$\dfrac{-7m}{2n}+\dfrac{m}{2n}$ $=\dfrac{-6m}{2n}=-\dfrac{3m}{n}$
3. Add $\dfrac{-3}{x^2-3x+2}+\dfrac{x+1}{x^2-3x+2}$. Answer: $\dfrac{x-2}{(x-2)(x-1)}$ $\dfrac{1}{x-1}$	$\dfrac{-3}{x^2-3x+2}+\dfrac{x+1}{x^2-3x+2}$ $\dfrac{-3(x+1)}{(x+1)(x-2)}=\dfrac{-3}{x-2}$

Example	Notes

4 & 5. Add.

$$\frac{x+3}{x^2-1}+\frac{x}{x^2-1}$$

$$\frac{2x+3}{(x+1)(x-1)}$$

$$\frac{x}{x+1}+\frac{1}{x+1}$$

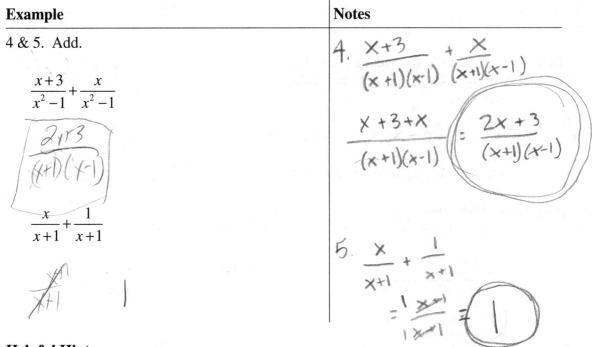

Helpful Hints

If rational expressions have a common denominator, they can be added in the same way as like fractions. For any rational expressions $\frac{a}{b}$ and $\frac{c}{b}$, $\frac{a}{b}+\frac{c}{b}=\frac{a+c}{b}$, where $b \neq 0$.

With all calculations with rational expressions, remember to simplify whenever possible by combining like terms, factoring, and dividing by common factors.

Concept Check

1. Are $\frac{5}{4x-7}$ and $\frac{-3}{(-7)+4x}$ like rational expression?

Practice

Add.

2. $\frac{5}{x-8}+\frac{3}{x-8}$

4. $\frac{3x^2-4x}{3x-7}+\frac{5x-14}{3x-7}$

3. $\frac{3u}{20q}+\frac{2u}{20q}$

5. $\frac{4}{x+5}+\frac{x+1}{x+5}$

Name: Sydney

Instructor: Schnabel

Date: 2/12/16

Section: _____

Adding and Subtracting Rational Expressions
Topic 27.2 Subtracting Like Rational Expressions

Vocabulary

rational expression • like rational expression • common denominator

1. A _rational expression_ can be written as a fraction of two algebraic expressions.

Step-by-Step Video Notes

Watch the Step-by-Step Video lesson and complete the examples below.

Example	Notes
1. Subtract $\dfrac{-2a}{3b} - \dfrac{5a}{3b}$. $\quad \dfrac{-7a}{3b}$ Subtract the numerators and keep the denominator the same. Answer:	$\dfrac{-2a}{3b} + \dfrac{-5a}{3b}$ $= \dfrac{-7a}{3b}$
2. Subtract $\dfrac{8x}{2x+3y} - \dfrac{3x}{2x+3y}$. Subtract the numerators and keep the denominator the same. $\dfrac{5x}{2x+3y}$ Answer:	$\dfrac{8x}{2x+3y} - \dfrac{-3x}{2x+3y}$ $= \dfrac{5x}{2x+3y}$
3. Subtract $\dfrac{3x^2+2x}{x^2-1} - \dfrac{10x-5}{x^2-1}$. $(3x^2+2x)(10x-5)$ Answer: $(3x-5)(x-1)$ $(x+1)(x-1)$	$\dfrac{3x^2+2x}{x^2-1} \quad \dfrac{-10x-5}{x^2-1}$ $\dfrac{(3x^2+2x)+(-10x+5)}{x^2-1}$ $\dfrac{3x^2-8x+5}{x^2-1}$ $3x^2-5x+3x+5$ $x(3x-5)\cdot 1(3x-5)$ $(x-1)(3x-5)$ $\dfrac{(3x-5)(x+1)}{(x+1)(x-1)} = \dfrac{3x-5}{x+1}$

Example	Notes

Example

4 & 5. Subtract.

$$\frac{3x}{x-4} - \frac{12}{x-4}$$

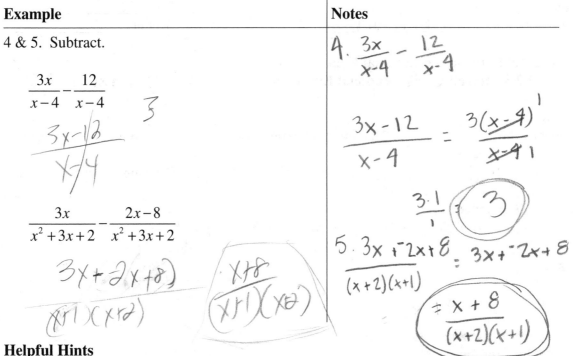

$$\frac{3x}{x^2+3x+2} - \frac{2x-8}{x^2+3x+2}$$

Notes

4. $\dfrac{3x}{x-4} - \dfrac{12}{x-4}$

$$\frac{3x-12}{x-4} = \frac{3(x-4)^1}{x-4^1}$$

$$\frac{3 \cdot 1}{1} = \boxed{3}$$

5. $\dfrac{3x + ^-2x + 8}{(x+2)(x+1)} = 3x + ^-2x + 8$

$$\boxed{\frac{=x+8}{(x+2)(x+1)}}$$

Helpful Hints

If rational expressions have the same denominator, they can be subtracted in the same way as fractions. For any rational expressions $\frac{a}{b}$ and $\frac{c}{b}$, $\frac{a}{b} - \frac{c}{b} = \frac{a-c}{b}$, where $b \neq 0$.

The numerator of the fraction being subtracted must be treated as a single quantity. Use parentheses when subtracting and be careful to use the correct signs.

Concept Check

1. Sophie subtracted $\dfrac{2x}{2x-3} - \dfrac{-3}{2x-3}$ and got an answer of 1. Is she correct? Explain.

Practice

Subtract.

2. $\dfrac{7}{5x-2} - \dfrac{8}{5x-2}$

4. $\dfrac{-7a}{4b} - \dfrac{a}{4b}$

3. $\dfrac{9m}{3m+n} - \dfrac{5m+7}{3m+n}$

5. $\dfrac{3x^2+17x}{9x+3} - \dfrac{x-5}{9x+3}$

Adding and Subtracting Rational Expressions
Topic 27.3 Finding the Least Common Denominator for Rational Expressions

Vocabulary

least common denominator (LCD)   •   like rational expression   •   factor

1. The _____LCD_____ of two or more rational expressions is the smallest expression that each of the denominators will divide into exactly.

Step-by-Step Video Notes

Watch the Step-by-Step Video lesson and complete the examples below.

Example	Notes
1. Find the LCD for $\dfrac{5}{2x-4}$, $\dfrac{6}{3x-6}$. Factor each denominator. $2(x+2)$ $3(x+2)$ List each different factor. $2^1 \cdot 3^1 (x-2)^1$ List each factor the greatest number of times it occurs in each denominator. Answer: $6(x-2)$	$\dfrac{5}{2x-4}$; $\dfrac{6}{3x-6}$ $2(x-2)$ $3(x-2)$ $2 \times 3 \times (x-2)$ $LCD =$ $\boxed{6(x-2)}$
2. Find the LCD for $\dfrac{5}{12ab^2c}$, $\dfrac{13}{18a^3bc^4}$. Answer: $36\,a^3bc^4$	$\dfrac{5}{12ab^2c}$ $\dfrac{13}{18a^3bc^4}$ $12 = 2 \cdot 2 \cdot 3 \cdot a \cdot b \cdot b \cdot c$ $18 = 2 \cdot 3 \cdot 3 \cdot a \cdot a \cdot a \cdot b \cdot c \cdot c \cdot c \cdot c$ LCD $= 2 \cdot 2 \cdot 3 \cdot 3 \cdot a \cdot a \cdot a \cdot b \cdot b \cdot c \cdot c \cdot c \cdot c$ $\boxed{= 36\, a^3 b^2 c^4}$

Example	Notes

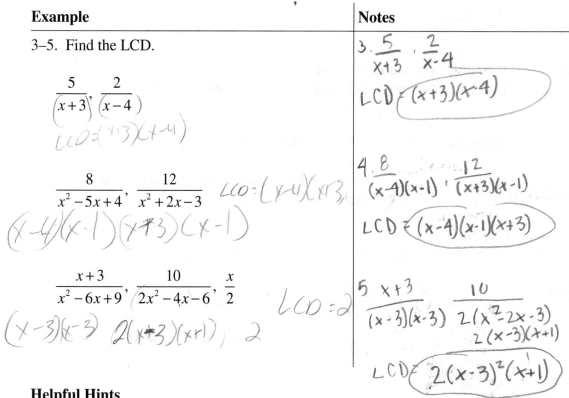

Example

3–5. Find the LCD.

$$\frac{5}{(x+3)}, \frac{2}{(x-4)}$$

$LCD = (x+3)(x-4)$

$$\frac{8}{x^2-5x+4}, \frac{12}{x^2+2x-3}$$

$LCD: (x-4)(x+3)$

$(x-4)(x-1)(x+3)(x-1)$

$$\frac{x+3}{x^2-6x+9}, \frac{10}{2x^2-4x-6}, \frac{x}{2}$$

$(x-3)(x-3) \quad 2(x+3)(x+1), \quad 2$

$LCD = 2$

Notes

3. $\frac{5}{x+3}, \frac{2}{x-4}$

$LCD = (x+3)(x-4)$

4. $\frac{8}{(x-4)(x-1)}, \frac{12}{(x+3)(x-1)}$

$LCD = (x-4)(x-1)(x+3)$

5. $\frac{x+3}{(x-3)(x-3)} \quad \frac{10}{2(x^2-2x-3)}$

$\quad 2(x-3)(x+1)$

$LCD = 2(x-3)^2(x+1)$

Helpful Hints

If a factor occurs more than once in any one denominator, the LCD will contain that factor repeated the greatest number of times that it occurs in any one denominator.

Be careful when lining up common factors. For example, x and $x-2$ are not common factors, but x and x^2 involve the same factor x, with the highest degree of 2.

Concept Check

1. Is $4x^4$ a common denominator of $\frac{5}{2x^3}$ and $\frac{y}{2x}$? Is it the LCD of $\frac{5}{2x^3}$ and $\frac{y}{2x}$? Explain.

Practice

Find the LCD.

2. $\frac{5}{9x+24}, \frac{11}{21x+56}$

4. $\frac{2}{x-3}, \frac{7}{x+6}$

3. $\frac{13}{30x^2y^3z}, \frac{16}{45x^3yz^4}$

5. $\frac{x-6}{x^3-4x^2+4x}, \frac{9x}{7x^2-21x+14}, \frac{4}{x}$

Adding and Subtracting Rational Expressions
Topic 27.4 Adding and Subtracting Unlike Rational Expressions

Vocabulary
equivalent rational expression • unlike rational expressions • like rational expressions

1. To add or subtract _____unlike rational expression_____, rewrite each rational expression as a(n)
   _____equivalent rational expression_____ whose denominator is the least common denominator.

Step-by-Step Video Notes
Watch the Step-by-Step Video lesson and complete the examples below.

Example	Notes
2. Add $\dfrac{5}{xy}+\dfrac{2}{y}$. Answer:	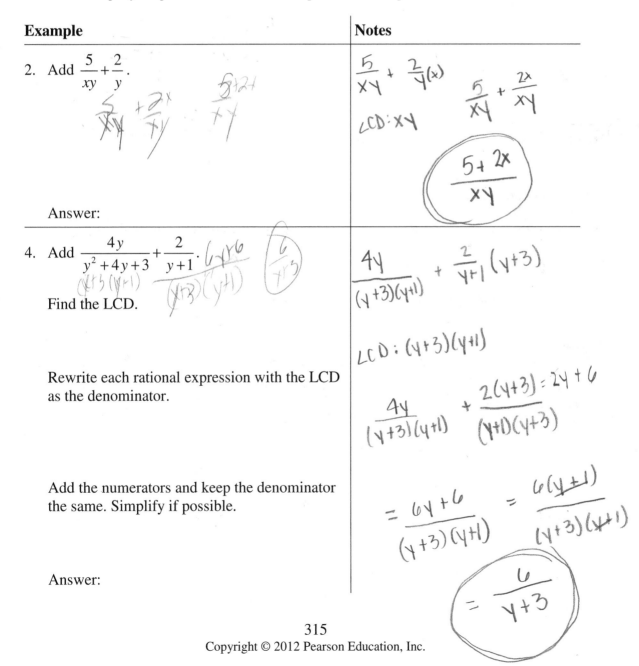
4. Add $\dfrac{4y}{y^2+4y+3}+\dfrac{2}{y+1}$. Find the LCD. Rewrite each rational expression with the LCD as the denominator. Add the numerators and keep the denominator the same. Simplify if possible. Answer:	

Example | **Notes**

5. Subtract $\dfrac{3x-4}{x-2} - \dfrac{5x-6}{2x-4}$.

[handwritten:] $\dfrac{6x-4+5x+6}{2x-4}$

$\dfrac{x+2}{2x-4}$ $\dfrac{x+2}{x-8}$

Answer: $\dfrac{1}{2}$

[handwritten Notes column:]
$\dfrac{3x-4}{x-2} + \dfrac{-5x+6}{2(x-2)}$

$LCD = 2, (x-2)$

$= \dfrac{6x-8+-5x+6}{2(x-2)}$

$\dfrac{x-2}{2(x-2)}$

$\boxed{=\dfrac{1}{2}}$

6. Subtract $\dfrac{-3}{x^2+8x+15} - \dfrac{1}{2x^2+7x+3}$.

[handwritten:] $(x+3)(x+5)$ $(2x+1)(x+3)$ $2x(x+3)$

$\dfrac{-6x-3+-x-5}{(2x+1)(x+3)(x+5)}$

Answer: $\dfrac{-7x-8}{(2x+1)(x+3)(x+5)}$

[handwritten Notes column:]
$\dfrac{-3}{(x+3)(x+5)} + \dfrac{-1}{(2x+1)(x+3)}$

$LCD = (x+3)(x+5)(2x+1)$

$\dfrac{-3(2x+1)}{(x+3)(x+5)(2x+1)} + \dfrac{-1(x+5)}{(x+3)(x+5)(2x+1)}$

$\dfrac{-6x-3}{} + \dfrac{-1x-5}{}$

$\boxed{=\dfrac{-7x-8}{(x+3)(x+5)(2x+1)}}$

Helpful Hints

It can be very easy to make a sign mistake when subtracting two rational expressions. You will find it helpful to place parentheses around the numerator of the second fraction so that you will not forget to subtract the entire numerator.

Remember that you can only add or subtract rational expressions with like denominators.

Concept Check

1. If $a \cdot b = c$, explain the steps you would use to add $\dfrac{x}{a} + \dfrac{x}{c}$.

Practice

Add.

2. $\dfrac{9}{m} + \dfrac{4}{mn}$

3. $\dfrac{3a-b}{a^2-9b^2} + \dfrac{4}{a+3b}$

Subtract.

4. $\dfrac{3x+9}{2x-10} - \dfrac{x-3}{x-5}$

5. $\dfrac{x}{x^2+3x-4} - \dfrac{x}{x^2+6x+8}$

Complex Rational Expressions and Rational Equations
Topic 28.1 Simplifying Complex Rational Expressions by Adding and Subtracting

Vocabulary

complex rational expression   •   least common denominator

1. A complex rational expression (also called a complex fraction) is a rational expression that contains a fraction in the numerator, in the denominator, or both.

Step-by-Step Video Notes
Watch the Step-by-Step Video lesson and complete the examples below.

Example	Notes

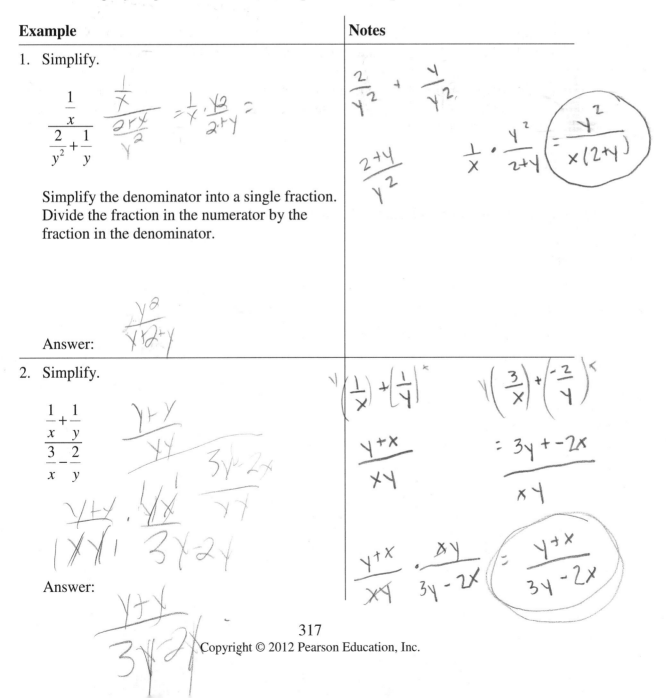

1. Simplify.

$$\dfrac{\dfrac{1}{x}}{\dfrac{2}{y^2}+\dfrac{1}{y}}$$

Simplify the denominator into a single fraction. Divide the fraction in the numerator by the fraction in the denominator.

Answer:

2. Simplify.

$$\dfrac{\dfrac{1}{x}+\dfrac{1}{y}}{\dfrac{3}{x}-\dfrac{2}{y}}$$

Answer:

Copyright © 2012 Pearson Education, Inc.

Example	Notes

4. Simplify.

$$\frac{\dfrac{3}{a+b} - \dfrac{3}{a-b}}{\dfrac{5}{a^2-b^2}}$$

Answer:

Notes (handwritten):

$$\frac{3(a-b)}{(a+b)(a-b)} - \frac{3(a+b)}{(a+b)(a-b)}$$

$$\frac{3a-3b-3a-3b}{(a+b)(a-b)} \cdot \frac{a^2-b^2}{5}$$

$$\frac{-6b}{(a+b)(a-b)} \cdot \frac{(a+b)(a-b)}{5} = \frac{-6b}{5}$$

Helpful Hints

The fraction bar in a complex fraction is both a grouping symbol and a symbol for division.

To simplify a complex rational expression by adding and subtracting, simplify the numerator and the denominator into single fractions as necessary. Divide the fraction in the numerator by the fraction in the denominator.

Concept Check

1. Write a multiplication problem that is equivalent to the complex fraction $\dfrac{\dfrac{4-x}{y}}{3+x}$.

Practice

Simplify.

2. $\dfrac{\dfrac{1}{a} + \dfrac{1}{a^2}}{\dfrac{2}{b^2}}$

4. $\dfrac{\dfrac{x}{x^2+4x+3} + \dfrac{2}{x+1}}{x+1}$

3. $\dfrac{\dfrac{1}{x} + \dfrac{1}{y}}{\dfrac{x}{2} - \dfrac{5}{y}}$

5. $\dfrac{\dfrac{6}{x^2-y^2}}{\dfrac{1}{x-y} + \dfrac{3}{x+y}}$

Complex Rational Expressions and Rational Equations
Topic 28.2 Simplifying Complex Rational Expressions by Multiplying by the LCD

Vocabulary
complex rational expressions   •   least common denominator (LCD)

1. One way to simplify a complex fraction is to find the ___LCD___ of each denominator in the complex fraction and multiply it by both the numerator and the denominator of the complex fraction.

Step-by-Step Video Notes
Watch the Step-by-Step Video lesson and complete the examples below.

Example	Notes
1. Simplify by multiplying by the LCD. $$\dfrac{\dfrac{3}{x}}{\dfrac{2}{x^2}+\dfrac{5}{x}}$$ Determine the LCD. Multiply the numerator and denominator by the LCD. Answer: $\dfrac{3x}{2+5x}$	$LCD = x^2$ $\dfrac{3x}{x^2}$ $\dfrac{2+5x}{x^2}$ $\dfrac{3x}{x^2} \cdot \dfrac{x^2}{2+5x}$ $= \dfrac{3x}{2+5x}$
2. Simplify by multiplying by the LCD. $$\dfrac{\dfrac{5}{ab^2}-\dfrac{2}{ab}}{3-\dfrac{5}{2a^2b}}$$ $\dfrac{2a(5-2b)}{b(6a^2b-5)}$ $\dfrac{2a(5-2b)}{b(6a^2b-5)}$ Answer:	$LCD: 2a^2b^2$ $\dfrac{2a(5-2b)}{b(6a^2b-5)}$ $\dfrac{5(2a^2b^2)}{ab^2} - \dfrac{2(2a^2b^2)}{ab}$ $10a \quad 3(2a^2b^2) \quad 4ab$ $\dfrac{10a-4ab}{6a^2b^2-5b}$ $\dfrac{5(2a^2b^2)}{2a^2b} \quad 5b$ $= \dfrac{5b}{6a^2b^2-5b}$

Example	Notes
3. Simplify by multiplying by the LCD. $$\frac{\frac{3}{a+b}-\frac{3}{a-b}}{\frac{5}{a^2-b^2}}$$ LCD: $(a+b)(a-b)$ $\frac{3a-3b-3a+3b}{5}$ Answer: $\frac{-60}{5}$	LCD: $(a+b)(a-b)$ $\frac{3(a+b)(a-b)}{a+b} = 3(a-b)$ $\frac{3(a-b)(a+b)}{a-b} = 3(a+b)$ $\frac{5(a+b)(a-b)}{(a+b)(a-b)} = 5$ $3a-3b-3a-3b = \frac{-6b}{5}$

Helpful Hints

To simplify a complex rational expression by multiplying by the LCD, you will often have to factor the denominators of the fractions in the expression to determine the LCD.

Concept Check

1. When is it easier to simplify a complex fraction by multiplying by the LCD?

Practice

Simplify by multiplying by the LCD.

2. $\dfrac{\frac{3}{a}+\frac{2}{b}}{\frac{5}{ab}}$

3. $\dfrac{\frac{3}{4x^2}-\frac{2}{y}}{\frac{7}{xy}-6}$

4. $\dfrac{\frac{8}{x^2-y^2}}{\frac{3}{x+y}+\frac{4}{x-y}}$

5. $\dfrac{\frac{2x}{x+3}+\frac{12}{x^2+8x+15}}{\frac{3}{x+5}}$

Complex Rational Expressions and Rational Equations
Topic 28.3 Solving Rational Equations

Vocabulary
rational expression   •   extraneous solution   •   apparent solution

1. A(n) _____ is an apparent solution that does not satisfy the original equation.

Step-by-Step Video Notes
Watch the Step-by-Step Video lesson and complete the examples below.

Example	Notes
1. Solve for x. $3\left(\dfrac{5}{x}+\dfrac{2}{3}=-\dfrac{3}{x}\right)\quad 3x$ Multiply each term of the equation by the LCD and solve the resulting equation. $15+2x=-9$ $-15\qquad\quad 15$ $2x=-24$ $x=-12$ Answer:	$\dfrac{5}{x}+\dfrac{2}{3}=\dfrac{-3}{x}\qquad -\dfrac{5}{12}+\dfrac{8}{12}=\dfrac{3}{12}$ $LCD:3x\qquad\qquad\qquad \dfrac{3}{12}=\dfrac{3}{12}\ \checkmark$ $\dfrac{5(3x)}{x}+\dfrac{2(3x)}{3}=\dfrac{-3(3x)}{x}$ $15+2x=-9$ $2x=-24$ $\boxed{x=-12}$
2. Solve for x. $\dfrac{6}{x+3}=\dfrac{3}{x}$ $6x=3x+9$ $-3x\quad 3x$ Answer: $x=3$	$\dfrac{6}{x+3}=\dfrac{3}{x}\qquad\quad \dfrac{6}{3+3}=\dfrac{3}{3}$ $\qquad\qquad\qquad\qquad \dfrac{6}{6}=\dfrac{3}{3}=1=1$ $LCD:x(x+3)$ $\dfrac{6(x(x+3))}{x+3}=\dfrac{3(x(x+3))}{x}$ $6x=3x+9$ $3x=9$ $\boxed{x=3}$

Example	Notes

3. Solve for x.

$$\frac{3}{x+5} - 1 = \frac{4-x}{2x+10}$$

(handwritten work)
$\frac{3}{-8+5} - 1 = \frac{4+8}{2(-8)\cdot10}$

$\frac{3}{-3}$

$-1-1 = -2$

$-2 = -2$ ✓

$6 - 2(x+5) = 4x$
$6 - 2x + 10 = 4 - x$
$-2x - 4 = 4 - x$

Answer: $-8 = x$

(Notes column handwritten)
$\frac{3}{x+5} - \frac{1}{1} = \frac{4-x}{2(x+5)}$

LCD: $2(x+5)$

$3 \cdot 2(x+5) - 1(2x+5) = \frac{4-x}{x(x+5)}$

$6 - 2(x+5) = 4 - x$

$6 \cdot 2x - 10 = 4$
$-8 = x$

---

4. Solve for y.

$$\frac{y}{y-2} - 4 = \frac{2}{y-2}$$

(handwritten work)
$y - 4(y-2) = 2$
$y - 4y + 8 = 2$
$-3y + 6 = 2$

Answer: $y = 2$ $\frac{-4}{3} = y$

$-4 = \frac{3y}{3}$

(Notes column handwritten)
LCD: $y-2$

$y - 4(y-2) = 2$
$y - 4y + 8 = 2$
$-3y = -6$
$y \neq 3$ ✱ no solutions

$3 - 4 = 2$
$-1 \neq 2$

$\frac{3}{3-2} - 4 = \frac{2}{3-2}$

Helpful Hints
In the case where a value makes a denominator in the equation zero, it is not a solution to the equation and therefore is not included in the domain.

Concept Check
1. Is 3 an extraneous solution to the equation $\dfrac{x}{x-3} + \dfrac{4}{5} = \dfrac{3}{x-3}$?

Practice
Solve for x.

2. $\dfrac{8}{x} + \dfrac{1}{2} = -\dfrac{2}{x}$

4. $\dfrac{2}{x-5} + 1 = \dfrac{3x-5}{4x-20}$

3. $\dfrac{9}{7x-4} = \dfrac{3}{2x}$

5. $\dfrac{2x}{x-3} = \dfrac{6}{x-3} + 3$

Name: _____   Date: _____

Instructor: _____   Section: _____

Complex Rational Expressions and Rational Equations
Topic 28.4 Direct Variation

Vocabulary
direct variation   •   constant of variation   •   function

1. If y varies directly as x, or y is directly proportional to x, then there is a positive constant k such that $y = kx$. In this case, the equation $y = kx$ is called an equation of

    _____.

2. The number k in the direct variation equation $y = kx$ is called the _____
    or the constant of proportionality.

Step-by-Step Video Notes
Watch the Step-by-Step Video lesson and complete the examples below.

Example	Notes
1 & 2. Suppose y varies directly as x. Find the constant of variation and the equation of direct variation for the following. $y = 12$ when $x = 3$ $y = -10$ when $x = -2$	

Example	Notes
3 & 4. Find the variation equation in which y varies directly as x and the following are true. Then find y for the given value of x. $y = -6$ when $x = -3$; $x = 10$ $y = 15$ when $x = 20$; $x = 8$	

Helpful Hints

In direct variation, as x increases or decreases, y also increases or decreases, respectively.

Concept Check

1. Suppose y varies directly as x, and $y = 8$ when $x = 2$. Is it possible to find value of x that corresponds to $y = 12$?

Practice

Suppose y varies directly as x. Find the constant of variation and the equation of direct variation for the following for x.

2. $y = 28$ when $x = 7$

Find the variation equation in which y varies directly as x and the following are true. Then find y for the given value of x.

4. $y = -42$ when $x = -6$; $x = 7$

3. $y = -24$ when $x = -8$

5. $y = 9$ when $x = 12$; $x = 8$

Complex Rational Expressions and Rational Equations
Topic 28.5 Inverse Variation

Vocabulary
inverse variation   •   constant of variation   •   function

1. If y varies inversely as x, or y is inversely proportional to x, then there is a positive constant k such that $y = \dfrac{k}{x}$. In this case, the equation $y = \dfrac{k}{x}$ is called an equation of

    _____.

2. The number k in the inverse variation equation $y = \dfrac{k}{x}$ is called the

    _____ or the constant of proportionality.

Step-by-Step Video Notes
Watch the Step-by-Step Video lesson and complete the examples below.

Example	Notes
1 & 2. Find an equation of variation in which y varies inversely as x for each of the following. $y = 5$ when $x = 2$ $y = -4$ when $x = -7$	

Example	Notes
3 & 4. Find an equation of variation in which y varies inversely as x and the following are true. Then find y for the given value of x. $y = -8$ when $x = -7$; $x = 4$ $y = 18$ when $x = \frac{1}{2}$; $x = 27$	

Helpful Hints

In inverse variation, as x increases or decreases, y decreases or increases, respectively.

Concept Check

1. Suppose y varies inversely as x, and $y = 3$ when $x = 1$. Can you find x given that $y = 1$?

Practice

Find an equation of variation in which y varies inversely as x for each of the following.

2. $y = 3$ when $x = 12$

Find an equation of variation in which y varies inversely as x and the following are true. Then find y for the given value of x.

4. $y = -12$ when $x = -4$; $x = -6$

3. $y = -11$ when $x = -2$

5. $y = 8$ when $x = 2.5$; $x = -2$

Name: _____ Date: _____

Instructor: _____ Section: _____

Complex Rational Expressions and Rational Equations
Topic 28.6 Applications

Vocabulary
inverse variation • constant of variation • direct variation • proportionality

1. The number k in the variation equations $y = kx$ and $y = \dfrac{k}{x}$ is called the

_____ or the constant of proportionality.

Step-by-Step Video Notes
Watch the Step-by-Step Video lesson and complete the examples below.

Example	Notes
1. The cost, C, of filling a tank of gas varies directly with the number of gallons of gas. If the cost of 1 gallon of gas is 2.50, find the cost of 10 gallons of gas. Let N represent the number of gallons of gas. Answer:	
2. The distance a spring stretches, d, varies directly with the weight of the object hung on the spring, w. If a 10-pound weight stretches a spring 6 inches, how far will a 35-pound weight stretch this spring? Answer:	

Example	Notes
4. If the voltage in an electric circuit is kept at the same level, the current I varies inversely with the resistance, R. The current measures 40 amperes when the resistance is 300 ohms. Find the current when the resistance is 100 ohms.	

Answer:

Helpful Hints
There are many applications that use direct and inverse variation. You will not always be told in a problem statement which kind of variation equation to use to solve a problem.

Concept Check
1. Does the amount of time it takes to drive to a certain destination vary directly or inversely with the average rate of speed you drive?

Practice
2. The amount of money you earn, S, varies directly with the number of hours that you work, t. If you earn $72 when you work 8 hours, determine how much you will earn if you work 13 hours.

3. The cost, C, of producing a certain tool varies inversely as the number produced, n. If 1000 of these tools are produced, the cost is $8 per unit. Find the cost per unit to produce 1600 tools.

4. If the temperature of a gas in a container is constant, the pressure P of the gas varies inversely with the volume V of the container. The pressure is 24 pounds per ft^2 when the volume is 4 ft^3. Find the pressure in a container (of the same temperature) with a volume of 6 ft^3.

Roots and Radicals
Topic 29.1 Square Roots

Vocabulary

square root   •   principal square root   •   negative square root

1. The _____ is one of two identical factors of a number.

2. The radical symbol $\sqrt{}$ is used to denote the _____ of a number.

Step-by-Step Video Notes
Watch the Step-by-Step Video lesson and complete the examples below.

Example	Notes
1–4. Find the square roots.	
$\sqrt{100}$ 10	
$\sqrt{\dfrac{9}{25}}$ $\dfrac{3}{5}$	
$-\sqrt{49}$ -7	
$\sqrt{0}$ 0	
7 & 8. Find the square roots.	
$\sqrt{49x^2}$ $7x$	
$\sqrt{(-6x)^2}$ $6x$	

Example	Notes
9 & 10. Find the square roots. $\sqrt{-144}$ No real number $-\sqrt{-9}$ No real number	

12. Approximate $\sqrt{75}$ by finding the two
   consecutive whole numbers that the square root
   lies between.

   Answer: $81, 64$ lies between $8, 9$

Helpful Hints
$8 < \sqrt{75} < 6$

If a variable appears in the radicand, assume it represents positive numbers only.

The square root of a negative number is not a real number.

Not every positive number has a rational square root. You can use a calculator to approximate the square roots of such numbers.

Concept Check
1. The value of $\sqrt{22}$ is between what two whole numbers?

Practice
Find the square roots.

2. $\sqrt{144}$

4. $\sqrt{64z^2}$

3. $-\sqrt{\dfrac{36}{121}}$

5. $\sqrt{-\dfrac{1}{4}}$

Roots and Radicals
Topic 29.2 Higher Roots

Vocabulary

square root   •   cube root   • n^{th} root

1. The _____ of a number is one of three identical factors.

2. The _____ is one of n identical factors of a number.

Step-by-Step Video Notes
Watch the Step-by-Step Video lesson and complete the examples below.

Example	Notes
1–3. Find the cube roots.	
$\sqrt[3]{27x^3}$ $3x$	
$-\sqrt[3]{\dfrac{8}{27}}$ $-\dfrac{2}{3}$	
$\sqrt[3]{-1}$ ~~no real number~~ -1	
4 & 5. Find the indicated roots.	
$\sqrt[5]{32}$ 2	
$\sqrt[4]{-81}$ no real number	

Example	Notes
6–8. Simplify the radicals. $\sqrt[5]{x^{10}}$ x^2 $\sqrt[4]{y^{12}}$ y^3 $\sqrt[4]{a^{24}b^{28}}$ $a^6 b^7$	

9 & 10. Simplify the radicals. $\sqrt[3]{-64x^6}$ $4x^2$ $\sqrt[4]{10,000x^4}$ $10x$	

Helpful Hints

A higher-order root is found using the radical sign $\sqrt[n]{a}$ where n is the index of the radical and a is called the radicand. To find an nth root, we can divide the exponent of the radicand by the index.

The cube root of a negative number is a negative number. If the index is even, the radicand must be nonnegative for the root to be a real number.

Concept Check

1. Find three different sets of whole number values for n and a if $\sqrt[n]{a} = 2$, and $n \geq 3$.

Practice

Simplify the radicals.

2. $\sqrt[3]{-27}$

3. $\sqrt[6]{x^{30}}$

4. Fill in the table below.

x	x^2	x^3	x^4	x^5
1	1	1	1	1
2	4		16	
3				243
4			256	
5		125		3125

Roots and Radicals
Topic 29.3 Simplifying Radicals

Vocabulary
square root   •   product rule for radicals   •   prime factorization

1. The _____ states that for all nonnegative real numbers a, b, and n, $\sqrt[n]{a} \cdot \sqrt[n]{b} = \sqrt[n]{ab}$ and $\sqrt[n]{ab} = \sqrt[n]{a} \cdot \sqrt[n]{b}$.

Step-by-Step Video Notes
Watch the Step-by-Step Video lesson and complete the examples below.

Example	Notes
1. Simplify the radical $\sqrt{50}$. Factor the radicand. If possible write the radicand as a product of a perfect square. $\sqrt{25} \cdot \sqrt{2}$ $5\sqrt{2}$ Use the Product Rule to separate the factors. Answer: $5\sqrt{2}$	
2 & 3. Simplify the radicals. $\sqrt{8}$ $\sqrt{4} \cdot \sqrt{2}$ $2\sqrt{2}$ $\sqrt{48}$ $\sqrt{8} \cdot \sqrt{6}$ $\sqrt{6} \cdot \sqrt{3}$ $4\sqrt{3}$	

Example	Notes

5 & 6. Simplify the radicals. Assume all variables represent nonnegative values. If the answer is not a real number, say so.

$\sqrt{x^5}$ $\quad x^{\frac{5}{2}}$ $\quad x^4 \cdot x^1$ $\quad x^{\frac{4}{2}}\sqrt{x}$ $\quad x^2\sqrt{x}$

$\qquad x^{5/2} = x^{2 \cdot 1}$ $\quad x^2\sqrt{x}$

$\sqrt[3]{y^{17}}$ $\quad y^{\frac{17}{3}}$ $\quad y^5 \sqrt[3]{y^2}$

7 & 8. Simplify the radicals. Assume all variables represent nonnegative values. If the answer is not a real number, say so.

$\sqrt{24x^3}$ $\quad \sqrt{4x^2}$ $\quad \sqrt{6x}$ $\quad 2x\sqrt{6x}$

$\sqrt[3]{-16x^{11}}$

$-2x^3 \sqrt[3]{2x^2}$ $\qquad -2x^3\sqrt[3]{2x^2}$

Helpful Hints

When simplifying, assume that all variables inside radicals represent nonnegative values.

When simplifying radicals, you can find the perfect root factors in a more difficult problem by looking at the prime factorizations of the radicands.

Concept Check

1. Of $\sqrt[3]{a^2}$, $\sqrt[3]{a^4}$, and $\sqrt[3]{-a^9}$, which expression cannot be simplified?

Practice

Simplify. Assume all variables represent nonnegative values.

2. $\sqrt{75}$
4. $\sqrt[3]{-24}$

3. $\sqrt{m^7}$
5. $\sqrt[3]{-250y^4}$

Name: _____   Date: _____

Instructor: _____   Section: _____

Roots and Radicals
Topic 29.4 Rational Exponents

Vocabulary
rational exponents   •   product rule for exponents   •   radical notation

1. _____ can be written as radicals.

2. The _____ states that $x^a \cdot x^b = x^{a+b}$.

Step-by-Step Video Notes
Watch the Step-by-Step Video lesson and complete the examples below.

Example	Notes
1 & 2. Write in radical notation. Simplify, if possible. $9^{1/2}$ $\sqrt{9}$ $\quad 3$ $x^{1/3}$ $\sqrt[3]{x}$	
3 & 4. Write in radical notation. Simplify, if possible. $6^{2/3}$ $\left(\sqrt[3]{6}\right)^2$ $\sqrt[3]{36}$ $x^{3/2}$ $\left(\sqrt[2]{x}\right)^3$ $\sqrt{x^3}$	

Example	Notes
5 & 6. Use radical notation to rewrite each expression. Simplify if possible. $\left(\dfrac{1}{9}\right)^{3/2}$ 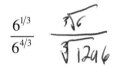 $\sqrt{\dfrac{\sqrt{1}}{\sqrt{9}}}$ $\left(\dfrac{1}{3}\right)^{3}$ $\boxed{\dfrac{1}{27}}$ $(3x)^{1/3}$ $\sqrt[3]{3x}$	
9. Use the rules of exponents to simplify. $\dfrac{6^{1/3}}{6^{4/3}}$ $\dfrac{\sqrt[3]{6}}{\sqrt[3]{1296}}$	

Helpful Hints

If m and n are integers greater than 1, with $\dfrac{m}{n}$ in simplest form, then $a^{m/n} = \sqrt[n]{a^{m}} = \left(\sqrt[n]{a}\right)^{m}$, as long as $\sqrt[n]{a}$ is a real number.

If an exponential expression has a negative rational exponent, write the expression with a positive exponent before evaluating the rational exponent.

Concept Check

1. Express $\sqrt[5]{32x^{10}}$ with rational exponents and simplify if possible.

Practice

Write in radical form. Simplify if possible.

2. $(4x)^{2/3}$

3. $8^{5/3}$

Simplify.

4. $x^{2/3} \cdot x^{1/4}$

5. $16^{-3/2}$

Roots and Radicals
Topic 29.5 The Pythagorean Theorem

Vocabulary

hypotenuse • leg • the Pythagorean Theorem

1. The side opposite the right angle in a right triangle is called the __hypotenus__.

2. In a right triangle, __the Pythagorian theorm__ states that the sum of the squares of the legs is equal to the square of the hypotenuse, or $leg^2 + leg^2 = hypotenuse^2$.

Step-by-Step Video Notes
Watch the Step-by-Step Video lesson and complete the examples below.

Example	Notes
1. Find the length of the hypotenuse. 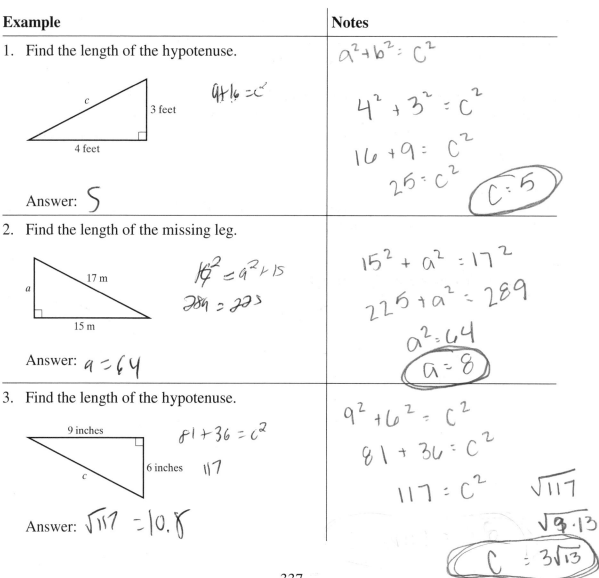 $9 + 16 = c^2$ Answer: 5	$a^2 + b^2 = c^2$ $4^2 + 3^2 = c^2$ $16 + 9 = c^2$ $25 = c^2$　$\boxed{c = 5}$
2. Find the length of the missing leg. 17 m · 15 m · a $17^2 = a^2 + 15$ $289 = 225$ Answer: $a = 64$	$15^2 + a^2 = 17^2$ $225 + a^2 = 289$ $a^2 = 64$ $\boxed{a = 8}$
3. Find the length of the hypotenuse. 9 inches · 6 inches · c $81 + 36 = c^2$　117 Answer: $\sqrt{117} = 10.8$	$9^2 + 6^2 = c^2$ $81 + 36 = c^2$ $117 = c^2$　$\sqrt{117}$ $\sqrt{9 \cdot 13}$ $\boxed{c = 3\sqrt{13}}$

Example	**Notes**

4 A slanted roof rises 5 feet vertically from the edge of the roof to the top. The roof covers 12 horizontal feet. How long is the slanted surface of the roof?

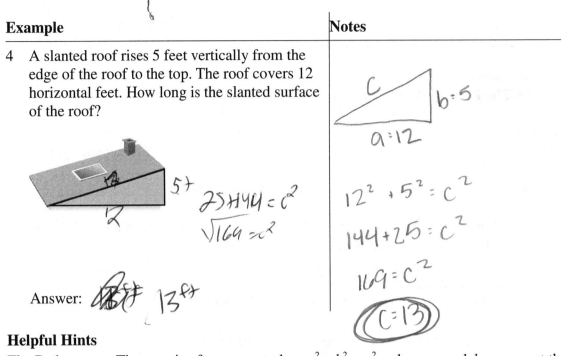

$25+44 = c^2$

$\sqrt{169} = c^2$

Answer: ~~16 ft~~ 13^{ft}

Notes column:

$b=5$

$a=12$

$12^2 + 5^2 = c^2$

$144 + 25 = c^2$

$169 = c^2$

$\boxed{C = 13}$

Helpful Hints

The Pythagorean Theorem is often presented as $a^2 + b^2 = c^2$, where a and b represent the legs of a right triangle and c represents the hypotenuse.

When using the Pythagorean Theorem, if the missing length is not a rational number, then express it either as a simplified radical or use a calculator to approximate it to a certain number of places. A given problem will usually specify whether an approximated or exact answer is required.

Concept Check

1. Why can't you always use the Pythagorean Theorem to find the missing side of a triangle with a side of length 2 m and a side of length 8 m?

Practice

Find the missing length in each right triangle.

2.   3.   4.

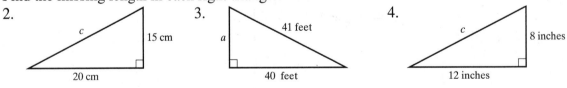

Roots and Radicals
Topic 29.6 The Distance Formula

Vocabulary
the Distance Formula   •   the Pythagorean Theorem

1. To calculate the distance between any two points (x_1, y_1) and (x_2, y_2) on a graph, use

   _____, which states $d = \sqrt{(x_2 - x_1)^2 + (y_2 - y_1)^2}$.

Step-by-Step Video Notes
Watch the Step-by-Step Video lesson and complete the examples below.

Example	Notes
1. Find the length of the line segment between the two points shown. Answer:	
3. Find the distance between $(-3, 2)$ and $(1, 5)$. Answer:	

Example	Notes
4 & 5. Use the Distance Formula to find the distance between the given points. $(6,-3)$ and $(1,9)$ $(2,1)$ and $(6,8)$	
6. Use the Distance Formula to find the distance between the points $(-13,-8)$ and $(-3,-3)$. Answer:	

Helpful Hints

Notice in example 3 that the horizontal and vertical distances between the two points make up the legs of a right triangle. The hypotenuse of this right triangle is the distance between the two points. This is how the Distance Formula is derived from the Pythagorean Theorem.

Be careful to avoid sign errors when finding the differences between the x- and y-coordinates of the points.

Concept Check

1. Will the Distance Formula also work for two points on the same horizontal line or the same vertical line? Explain.

Practice

Find the distance between the two points shown on the graph.

2.

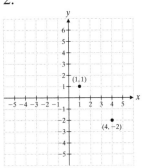

Use the Distance Formula to find the distance between the given points.

3. $(-3,-4)$ and $(5,11)$

4. $(4,7)$ and $(8,13)$

Name: _Sydney_   Date: _3/7/16_

Instructor: _Schnabel_   Section: _____

Operations of Radical Expressions

Topic 30.1 Adding and Subtracting Radical Expressions

Vocabulary

like radicals   •   like terms   •   radical expressions

1. Radicals with the same radicand and the same index are called _like radicals_.

Step-by-Step Video Notes

Watch the Step-by-Step Video lesson and complete the examples below.

Example	Notes
1–3. Combine like terms. $7x+9x$ 16y $13xy^2+11x^2y$ $5x^{1/2}+2x^{1/2}$ $7x^{1/2}$ $7\sqrt{x}$	$7x+9x=\boxed{16x}$ $13xy^2+11x^2y$ $=$ not like terms $5x^{1/2}+2x^{1/2}$ $=7x^{1/2}\boxed{=7\sqrt{x}}$
4–6. Simplify, if possible. Assume that all variables are nonnegative real numbers. $5\sqrt{3}+7\sqrt{3}$ $12\sqrt{3}$ $3\sqrt{2}-9\sqrt{2}$ -6 $-6\sqrt{xy}+2\sqrt[3]{xy}$	$5\sqrt{3}+7\sqrt{3}\boxed{=12\sqrt{3}}$ $3\sqrt{2}-9\sqrt{2}\boxed{=-6\sqrt{2}}$ $-6\sqrt{xy}+2\sqrt[3]{xy}$ $=$ not like radicals!

Example	Notes
8. Simplify. $5\sqrt{3} - \sqrt{27} + 2\sqrt{32}$ *(handwritten work)* $5\sqrt{3} - 3\sqrt{3} + 4\sqrt{2} \cdot 2$ *(handwritten)* $2\sqrt{3} + 4\sqrt{2}$	$5\sqrt{3} - \sqrt{9 \cdot 3} + 2\sqrt{16 \cdot 2}$ $5\sqrt{3} - 3\sqrt{3} + 8\sqrt{2}$ $2\sqrt{3} + 8\sqrt{2}$
10. Simplify. Assume that all variables are nonnegative real numbers. $3x\sqrt[3]{54x^4} - 3\sqrt[3]{16x^7}$ *(handwritten work)* $9x^2\sqrt[3]{2x} - 6x^2\sqrt[3]{2x}$ $3x\sqrt[3]{2x}$	$3x\sqrt[3]{54x^4} - 3\sqrt[3]{16x^7}$ $3x\sqrt[3]{27 \cdot 2x^4} - 3\sqrt[3]{8 \cdot 2x^7}$ $3 \cdot 3x\sqrt[3]{2x^4} - 3 \cdot 2$ $x^{\frac{7}{3}}$ $-6x^2\sqrt[3]{2x}$ $9x\sqrt[3]{2 \cdot x^{\frac{4}{3}}}$ $9x^2\sqrt[3]{2x} - 6x^2\sqrt[3]{2x}$ $= 3x^2\sqrt[3]{2x}$

Helpful Hints

A radical is written using the radical sign $\sqrt[n]{a}$ where n is the index of the radical and a is called the radicand.

For all nonnegative real numbers a, b, and n, where n is an integer greater than 1, $\sqrt[n]{ab} = \sqrt[n]{a} \cdot \sqrt[n]{b}$.

Concept Check

1. Are $x\sqrt{7}$ and $7\sqrt{x}$ like radicals? Explain why or why not.

Practice

Simplify. Assume variables represent nonnegative numbers.

2. $11\sqrt{7} - 12\sqrt{7}$

3. $\sqrt{48} + \sqrt{75}$

4. $-\sqrt{12} + 6\sqrt{27} - 4\sqrt{28}$

5. $\sqrt{72x} - 2x\sqrt{3} - 5\sqrt{2x} + x\sqrt{27}$

Name: _Sydney_   Date: _2/7/14_

Instructor: _Schnabel_   Section: _____

Operations of Radical Expressions
Topic 30.2 Multiplying Radical Expressions

Vocabulary
Distributive Property   •   Product Rule for Radicals   •   FOIL method

1. The _Product rule for radicals_ states that for all nonnegative real numbers a, b, and n, where n is an integer greater than 1, $\sqrt[n]{a} \cdot \sqrt[n]{b} = \sqrt[n]{ab}$.

Step-by-Step Video Notes
Watch the Step-by-Step Video lesson and complete the examples below.

Example	Notes
2. Multiply. $$\sqrt{5} \cdot \sqrt{3} \quad \sqrt{15}$$ Use the Product Rule for radicals. Answer: $\sqrt{15}$	$\sqrt{5} \cdot \sqrt{3} =$ $$\boxed{\sqrt{15}}$$
4. Multiply. $$\left(\sqrt{12}\right)\left(-5\sqrt{3}\right)$$ $$-5\sqrt{36}$$ Use the Product Rule for radicals. Simplify, if possible. Answer: $-5\sqrt{36}$ -30	$$\left(\sqrt{12}\right)\left(-5\sqrt{3}\right)$$ $$= -5\sqrt{36}$$ $$= -5 \cdot 4$$ $$\boxed{= -30}$$

343

$$(a+b)^2 = a^2 + 2ab + b^2 \qquad (a-b)^2 = a^2 - 2ab + b^2$$

Example	**Notes**
6. Multiply. $$\left(\sqrt{2}+3\sqrt{5}\right)\left(2\sqrt{2}-\sqrt{5}\right)$$ $2\sqrt{4}-\sqrt{10}+6\sqrt{10}-3\sqrt{25}$ $11+5\sqrt{10}$ Answer: $-11+5\sqrt{10}-3\sqrt{25}$	$$\left(\sqrt{2}+3\sqrt{5}\right)\left(2\sqrt{2}-\sqrt{5}\right)$$ $2\sqrt{4}-\sqrt{10}+6\sqrt{10}-3\sqrt{25}$ $=2\sqrt{4}+5\sqrt{10}-3\sqrt{25}$ $2\cdot2 \;+5\sqrt{10}\;-3\cdot5$ $=-11+5\sqrt{10}$
8. Simplify. Assume all variables represent nonnegative numbers. $$\left(\sqrt{7}+\sqrt{3x}\right)^2$$ $\sqrt{49}\;\sqrt{21x}+\sqrt{21x}+\sqrt{9x^2}$ $7+\sqrt{42x}+3x$ Answer: $2\sqrt{21x}$ $\;\;3x+\sqrt{2}\,\sqrt{21x}+7$	$$\left(\sqrt{7}+\sqrt{3x}\right)^2$$ $$\left(\sqrt{7}+\sqrt{3x}\right)\left(\sqrt{7}+\sqrt{3x}\right)$$ $\sqrt{49}\;+\sqrt{21x}+\sqrt{21x}\;+\sqrt{9x^2}$ $7+2\sqrt{21x}+3x$

Helpful Hints
Apply multiplicative properties like the Distributive Property, the Product Rules, the FOIL method, the difference of two squares and squaring binomials when multiplying radical expressions, just as you would with other types of expressions.

Recall that $\left(\sqrt{x}\right)^2 = x$.

Concept Check
1. Addison claims that $\left(\sqrt{-3}\right)\left(\sqrt{-3}\right)=\left(\sqrt{9}\right)=3$. Is she correct? Explain why or why not.

Practice
Multiply. Assume all variables represent nonnegative numbers.

2. $\left(4x\sqrt{2y}\right)\left(7x\sqrt{8}\right)$

4. $\left(4\sqrt{5}+3\sqrt{2}\right)\left(4\sqrt{5}-3\sqrt{2}\right)$

3. $\sqrt{3x}\left(2\sqrt{6x}+7\sqrt{12}\right)$

5. $\sqrt[3]{7x}\left(\sqrt[3]{49x^2}-5\sqrt[3]{2x}\right)$

Name: _Sydney_   Date: _3/9/16_

Instructor: _Schnabel_   Section: _____

Operations of Radical Expressions
Topic 30.3 Dividing Radical Expressions

Vocabulary

Quotient Rule for Radicals   •   radical expressions

1. The ___Quotient rule for radicals___ states that for all nonnegative real numbers a, b, and n, where n is an integer greater than 1, and $b \neq 0$, $\dfrac{\sqrt[n]{a}}{\sqrt[n]{b}} = \sqrt[n]{\dfrac{a}{b}}$.

Step-by-Step Video Notes

Watch the Step-by-Step Video lesson and complete the examples below.

Example	Notes
1. Divide. $\dfrac{\sqrt{75}}{\sqrt{3}} = \sqrt{\dfrac{75}{3}} \quad \sqrt{25} = 5$	$\dfrac{\sqrt{75}}{\sqrt{3}} = \sqrt{\dfrac{75}{3}} = \sqrt{25}$ $=5$
Answer:	
2 & 3. Divide. $\dfrac{\sqrt{9}}{\sqrt{16}} \qquad \dfrac{3}{4}$	$\dfrac{\sqrt{9}}{\sqrt{14}} = \dfrac{3}{4}$
$\dfrac{\sqrt{72}}{\sqrt{8}} \quad \dfrac{\sqrt{8}\sqrt{9}}{\sqrt{8}} = 3$	$\dfrac{\sqrt{72}}{\sqrt{8}} = \dfrac{\sqrt{8}\cdot 9}{\sqrt{8}} = \sqrt{9} = 3$

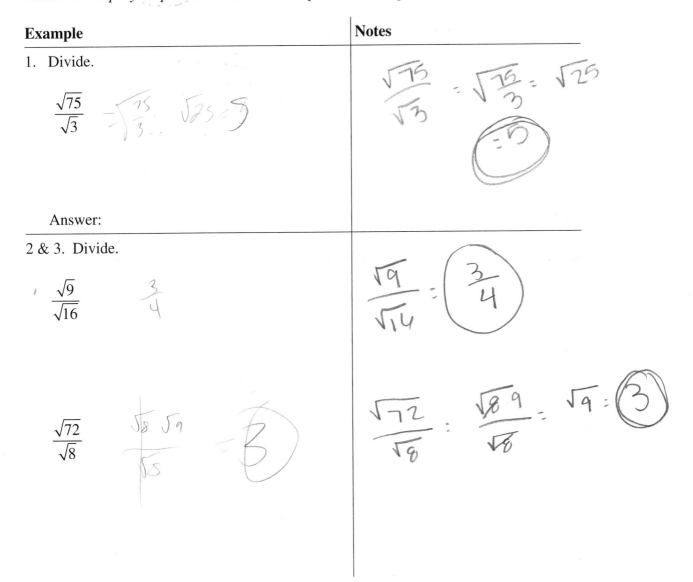

345
Copyright © 2012 Pearson Education, Inc.

Example	Notes
4 & 5. Divide.	

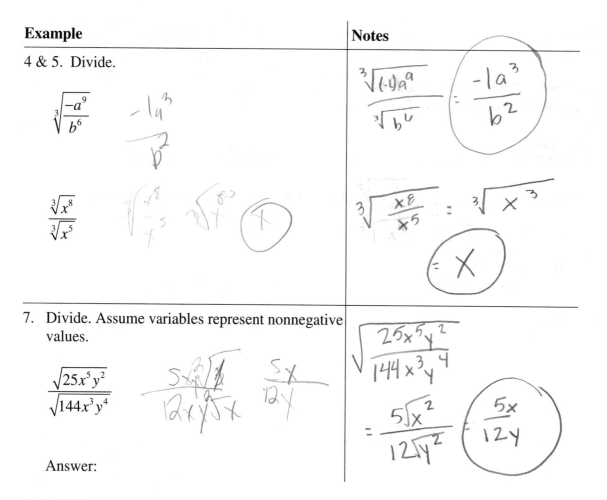

$$\sqrt[3]{\dfrac{-a^9}{b^6}}$$

$$\dfrac{\sqrt[3]{x^8}}{\sqrt[3]{x^5}}$$

7. Divide. Assume variables represent nonnegative values.

$$\dfrac{\sqrt{25x^5y^2}}{\sqrt{144x^3y^4}}$$

Answer:

Helpful Hints
The Quotient Rule for Radicals can be very flexible. You can simplify the numerators and/or denominators first, or you can divide the radicands first, depending on the situation.

Concept Check
1. Simplify Example 5 a different way than you did originally. Do you get the same answer?

Practice
Divide. Assume all variables represent nonnegative numbers.

2. $\dfrac{\sqrt{150}}{\sqrt{6}}$

4. $\dfrac{\sqrt{25a^6}}{\sqrt{81b^{12}c^4}}$

3. $\dfrac{\sqrt{63a^9b^7}}{\sqrt{7a^3b}}$

5. $\sqrt{\dfrac{147a^2b^6}{3a^8b^4}}$

Operations of Radical Expressions
Topic 30.4 Rationalizing the Denominator

Vocabulary
rationalizing the denominator • Identity Property of Multiplication • conjugates

1. Performing operations to remove a radical from a denominator is called
   <u>rationalizing the denominator</u>

Step-by-Step Video Notes
Watch the Step-by-Step Video lesson and complete the examples below.

Example	Notes
1. Simplify by rationalizing the denominator. $\dfrac{1}{\sqrt{2}}$	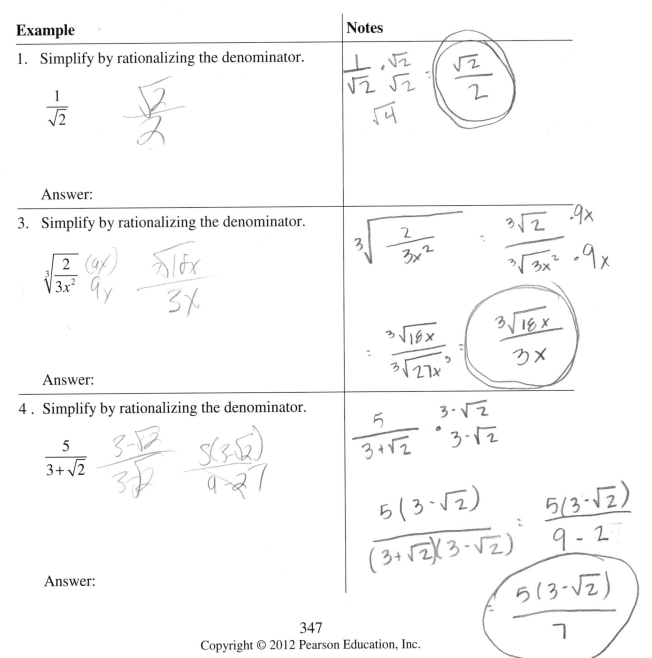 $\dfrac{1}{\sqrt{2}} \cdot \dfrac{\sqrt{2}}{\sqrt{2}} = \boxed{\dfrac{\sqrt{2}}{2}}$ $\sqrt{4}$
Answer:	
3. Simplify by rationalizing the denominator. $\sqrt[3]{\dfrac{2}{3x^2}}$ $\dfrac{\sqrt[3]{18x}}{3x}$	$\sqrt[3]{\dfrac{2}{3x^2}} = \dfrac{\sqrt[3]{2}}{\sqrt[3]{3x^2}} \cdot \dfrac{9x}{9x}$ $= \dfrac{\sqrt[3]{18x}}{\sqrt[3]{27x^3}} = \boxed{\dfrac{\sqrt[3]{18x}}{3x}}$
Answer:	
4. Simplify by rationalizing the denominator. $\dfrac{5}{3+\sqrt{2}}$ $\dfrac{3-\sqrt{2}}{3\sqrt{2}}$ $\dfrac{5(3-\sqrt{2})}{9-2?}$	$\dfrac{5}{3+\sqrt{2}} \cdot \dfrac{3-\sqrt{2}}{3-\sqrt{2}}$ $\dfrac{5(3-\sqrt{2})}{(3+\sqrt{2})(3-\sqrt{2})} = \dfrac{5(3-\sqrt{2})}{9-2}$ $\boxed{\dfrac{5(3-\sqrt{2})}{7}}$
Answer:	

347
Copyright © 2012 Pearson Education, Inc.

Example	Notes

5. Simplify by rationalizing the denominator.

$$\frac{\sqrt{7}+\sqrt{3}}{\sqrt{7}-\sqrt{3}}$$

Answer:

Helpful Hints

A rational expression is not considered to be in simplest form if there is an irrational expression in the denominator.

When rationalizing a denominator, you can simplify the radical in the denominator first, and then rationalize the denominator, or you can rationalize first, and then simplify the fraction.

The product of two conjugates is always rational.

Concept Check

1. Explain why $\dfrac{2}{1-\sqrt{2}}$ cannot be rationalized by multiplying by $\dfrac{\sqrt{2}}{\sqrt{2}}$.

Practice

Simplify by rationalizing the denominator. Assume variables represent nonnegative numbers.

2. $\dfrac{2x}{\sqrt{7}}$

4. $\sqrt[3]{\dfrac{4}{3x^2}}$

3. $\dfrac{5}{\sqrt{75}}$

5. $\dfrac{8}{3-\sqrt{5}}$

Name: _Sydney_   Date: _3/9/16_

Instructor: _____Schnabel_____   Section: _____

Operations of Radical Expressions
Topic 30.5 Solving Radical Equations

Vocabulary

Squaring Property of Equality • radical equations • reverse operations

1. The ___squaring property of equality___ states that for all real numbers a, and b, if $a = b$,
   then $a^2 = b^2$.

Step-by-Step Video Notes

Watch the Step-by-Step Video lesson and complete the examples below.

Example	Notes
1. Solve. $\sqrt{x} = 7$ Square both sides of the equation. Check. $x = 49$ Answer:	$(\sqrt{x})^2 = (7)^2$ $\sqrt{49} = 7$ $7 = 7$ ✓ $\boxed{x = 49}$
3. Solve $3 + \sqrt{x} = 7$. Perform operations to get the radical by itself on one side of the equation. Square both sides of the equation and solve. Answer: 16	$3 + \sqrt{x} = 7$ $(\sqrt{x})^2 = (4)^2$ $\sqrt{16} = 4$ $4 = 4$ ✓ $\boxed{x = 16}$ $3 + \sqrt{16} = 7$ $3 + 4 = 7$ ✓

Example	Notes

4. Solve.

$$\left(\sqrt{x-5}\right)^2 = 12^2$$

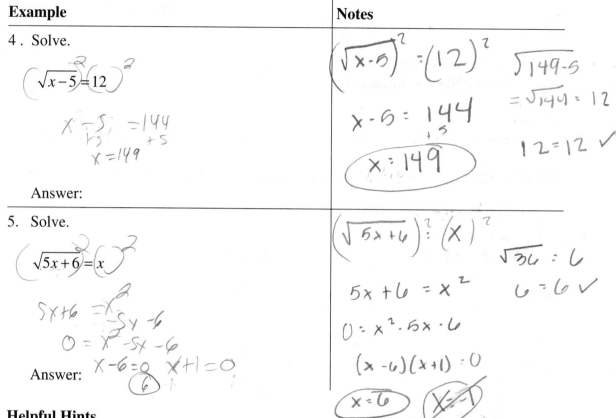

$x - 5 = 144$

$x = 149$

Answer:

---

5. Solve.

$$\left(\sqrt{5x+6}\right)^2 = x^3$$

$5x + 6 = x^2$

$0 = x^2 - 5x - 6$

$x - 6 = 0 \quad x + 1 = 0$

$6 \quad -1$

Answer:

Notes column:

$$\left(\sqrt{5x+6}\right)^2 = (x)^2$$

$\sqrt{36} = 6$

$5x + 6 = x^2$

$6 = 6 \checkmark$

$0 = x^2 - 5x - 6$

$(x - 6)(x + 1) = 0$

$x = 6 \qquad \cancel{x = -1}$

(Example 4 notes): $\sqrt{149-5}$ $= \sqrt{144} = 12$ $12 = 12 \checkmark$

Helpful Hints

Just as you can apply the four basic operations to both sides of an equation, squaring both sides of an equation will result in an equivalent equation.

Check all possible solutions to make sure they work in the original equation. Solutions to radical equations must be verified.

Concept Check

1. Is 64 a solution to $-\sqrt{x} = -8$? Is 64 a solution to $\sqrt{-x} = -8$? Explain why or why not for both cases.

Practice

Solve.

2. $-4 + \sqrt{x} = 5$

3. $\sqrt{2x-5} = 3$

4. $\sqrt{3x+4} = x$

5. $\sqrt{x+6} + 8 = 14$

Solving Quadratic Equations
Topic 31.1 Introduction to Solving Quadratic Equations

Vocabulary

quadratic equation   •   solution   •   parabola   • x-intercept

1. A(n) __quadratic equation__ is an equation of the form $ax^2 + bx + c = 0$ where $a \neq 0$.

2. A(n) __x-intercept__ of the graph of an equation is a point where the curve crosses the x-axis, or where the value of y is 0.

Step-by-Step Video Notes

Watch the Step-by-Step Video lesson and complete the examples below.

Example	Notes
1–3. Write each quadratic equation in standard form.	1. $7x^2 = 6x + 8$ $\boxed{7x^2 - 5x - 8 = 0}$
$7x^2 = 5x + 8$ $7x^2 - 5x - 8 = 0$	
$6x - 2x^2 = -3$ $0 = 2x^2 - 6x - 3$	2. $6x - 2x^2 = -3$ $\boxed{0 = 2x^2 - 6x - 3}$
$4x^2 = 9$ $4x^2 - 9 = 0$	3. $4x^2 = 9$ $\boxed{4x^2 - 9 = 0}$
4–6. Determine the number of solutions for each quadratic equation graphed below.	4. 2 solutions 5. 1 solutions 6. 0 solutions

Copyright © 2012 Pearson Education, Inc.

Example	**Notes**

7 & 8. Find the real solution(s) of each quadratic equation, if any exist, by using the graph of the equation.

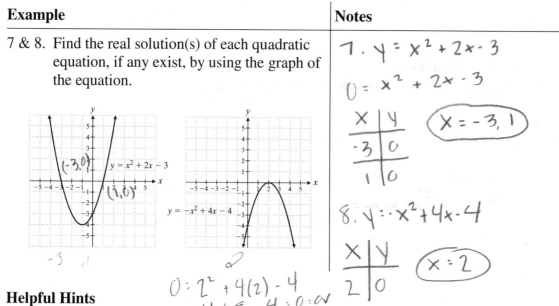

$y = x^2 + 2x - 3$

$y = -x^2 + 4x - 4$

Notes (handwritten):

7. $y = x^2 + 2x - 3$

$0 = x^2 + 2x - 3$

x	y
-3	0
1	0

$\boxed{x = -3, 1}$

8. $y = -x^2 + 4x - 4$

x	y
2	0

$\boxed{x = 2}$

$0 = 2^2 + 4(2) - 4$

$-4 + 8 - 4 = 0 = 0 \checkmark$

Helpful Hints

The graph of a quadratic equation $y = ax^2 + bx + c$ is called a parabola.

A quadratic equation may have two real solutions, one real solution, or no real solutions.

Concept Check

1. How can you use the graph of a quadratic equation to determine the real solutions of the equation?

Practice

Write each quadratic equation in standard form.

2. $x^2 + 9 = -6x$

3. $2x + 3 = 5x^2$

Find the real solution(s) of each quadratic equation by using the graph of the equation.

4.

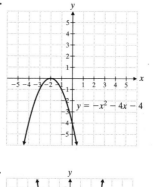

$y = -x^2 - 4x - 4$

5.

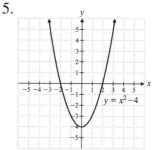

$y = x^2 - 4$

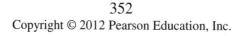

Name: _Sydney_

Instructor: _Schnabl_

Date: _3/21/16_

Section: _____

Solving Quadratic Equations
Topic 31.2 Solving Quadratic Equations by Factoring

Vocabulary

standard form • factor • quadratic expression

1. To solve a quadratic equation by factoring, first make sure the equation is in the
standard form, $ax^2 + bx + c = 0$.

Step-by-Step Video Notes

Watch the Step-by-Step Video lesson and complete the examples below.

Example	Notes
1. Solve $x^2 - 9x = 0$ by factoring.	$x^2 - 9x = 0$
Factor completely. Set each factor equal to zero.	$x(x-9) = 0$
$0 \quad 9$ $\quad x = 0 \quad x-9=0$ $\quad x=9$ $x(x-9)$	$x = 0 \qquad x-9=0$ $\qquad\qquad x=9$
Solve each equation. Check each solution. $0^2 + 0 - 0 \qquad 81 - 81 = 0$	$(0)^2 - 9(0) = 0$ $\qquad 0 \cdot 0 = 0$ $\qquad 0 = 0 \checkmark$ $9^2 - 9(9) = 0$ $81 - 81 = 0$ $\qquad 0 = 0 \checkmark$
Answer: $0, 9$	
2. Solve $5x^2 - 14x - 3 = 0$ by factoring. $(5x+1)(x-3) = 0$ $5x+1=0 \quad x-3=0$ $x = -\frac{1}{5}$ $\qquad x = 3$	$\begin{matrix} -15 \\ 1 \diagdown -15 \\ -14 \end{matrix}$ $\quad (5x^2 + x)(15x - 3)$ $x(5x+1) \cdot 3(5x+3)$ $(x-3)(5x+1)$ $x-3=0 \qquad 5x+1=0$ $x = 3 \qquad\quad 5x = -1$ $\qquad\qquad\qquad x = -\frac{1}{5}$ $5(\tfrac{1}{5})^2 - 14(-\tfrac{1}{5})_3$ $\frac{1}{5} + \frac{14}{5} - \frac{15}{5}$ $= 0$ $0 = 0 \checkmark$
Answer:	$5(3)^2 - 14(3) - 3 = 0$ $45 - 43 - 3 = 0$ $0 = 0 \checkmark$

353

Example	Notes
3. Solve $9x^2 = 24x - 15$ by factoring. $9x^2 \cdot 24/15 = 0$ $3(3x^2 - 8y + 5) = 0$ $25 = 25$ ✓ $3(3x-5)(x-1) = 0$ $9 = 9$ ✓ $3x-5 = 0$ $x-1 = 0$ Answer: $1, \frac{5}{3}$	$9x^2 - 24x + 15 = 0$ $3(3x^2 - 8x + 5) = 0$ $3(3x-5)(x-1) = 0$ $3x - 5 = 0$ ⟨$x = 1$⟩ $3x = 5$ ⟨$x = \frac{5}{3}$⟩
4. Solve $x^2 - 2x = -1$ by factoring. $x^2 - 2x + 1 = 0$ $(x-1)(x-1) = 0$ $-1 = -1$ ✓ $x - 1 = 0$ Answer: 1	$x^2 - 2x + 1 = 0$ $(x-1)(x-1) = 0$ $x - 1 = 0$ ⟨$x = 1$⟩

Helpful Hints

When solving a quadratic equation by factoring, set each factor containing a variable equal to zero. Remember, if you factor out the variable or a monomial term containing the variable as the GCF, then 0 is one of the solutions of the equation.

Concept Check

1. Rewrite $y = (x-7)^2$ by setting each factor equal to zero to solve. How many solutions does this equation have?

Practice

Solve by factoring.

2. $3x^2 + 15x = 0$

4. $7x^2 + 84 = 56x$

3. $5x^2 + 3x - 14 = 0$

5. $x^2 + 10x + 25 = 0$

Solving Quadratic Equations
Topic 31.3 Solving Quadratic Equations by Using the Square Root Property

Vocabulary
Square Root Property   •   quadratic equation

1. A _quadratic equation_ is an equation of the form $ax^2 + bx + c = 0$ where $a \neq 0$.

Step-by-Step Video Notes
Watch the Step-by-Step Video lesson and complete the examples below.

Example	Notes
2. Solve $x^2 = 48$ for x. Use the Square Root Property. $x = \sqrt{48}$ $\sqrt{16}\sqrt{3}$ $x = 4\sqrt{3}$ $x = 4\sqrt{3}$ $(-4\sqrt{3})^2$ $(4\sqrt{3})^2$ Simplify and check your solutions. Answer: $4\sqrt{3}$ $-4\sqrt{3}$	$\sqrt{x^2} = \sqrt{48}$ $x = \sqrt{48}$ $x = \sqrt{4 \cdot 12}$ $x = 2\sqrt{4 \cdot 3}$ $x = 4\sqrt{3}$ $\boxed{x = \pm 4\sqrt{3}}$
3. Solve $x^2 = -4$ for x. Answer: no solutions	$\sqrt{x^2} = \sqrt{-4}$ no solution!

Example	Notes
4. Solve $3x^2 + 2 = 77$ for x. $3x + 2 = 77$ $\dfrac{3x^2}{3} = \dfrac{75}{3} \quad \sqrt{x^2} = \sqrt{25}$ $x = -$ Answer: $5, -5$	$3x^2 + 2 = 77$ $\dfrac{3x^2}{3} = \dfrac{75}{3}$ $\sqrt{x^2} = \sqrt{25} \qquad x = \pm\sqrt{25}$ $\boxed{x = \pm 5}$
5. Solve $(4x-1)^2 = 5$ for x. $4x - 1 = \sqrt{5}$ $4x = \sqrt{5} \qquad \dfrac{1\sqrt{5}}{4}$ $x = \dfrac{1+\sqrt{5}}{4} \quad x = \dfrac{1-\sqrt{5}}{4}$ Answer:	$(4x-1)^2 = 5$ $4x - 1 = \pm\sqrt{5}$ $\dfrac{4x}{4} = \dfrac{1 \pm \sqrt{5}}{4}$ $\boxed{x = \dfrac{1 \pm \sqrt{5}}{4}}$

Helpful Hints

The notation $\pm\sqrt{a}$ is a shorthand way of writing "$+\sqrt{a}$ or $-\sqrt{a}$".

The Square Root Property can only be used to solve an equation of the form $x^2 = a$ if $a \geq 0$. If $a < 0$, the equation has no real solutions.

Concept Check

1. Mark uses the square root property to solve the equation $x^2 + 4 = 0$ and incorrectly gets $x = \pm 2$ as his answer. Explain his error and state the correct answer.

Practice

Solve for x.

2. $x^2 = 28$

3. $x^2 = 196$

4. $6x^2 + 2 = 98$

5. $(2x-3)^2 = 25$

Solving Quadratic Equations
Topic 31.4 Solving Quadratic Equations by Completing the Square

✓

Vocabulary

completing the square   •   the Square Root Property   •   perfect square trinomial

1. To solve an equation like $x^2 + bx = c$, we can add a constant to both sides of the equation so that the left side becomes a perfect square trinomial. This method is called ___completing the square___

Step-by-Step Video Notes

Watch the Step-by-Step Video lesson and complete the examples below.

Example	Notes
1 & 2. Fill in the blanks to create a perfect square trinomial. $$x^2 + 8x + \underline{16} = \left(x + \underline{4}\right)^2$$ $$x^2 - 12x + \underline{?} = \left(x - \underline{6}\right)^2$$	1. $x^2 + 8x + \underline{16} = (x + \underline{4})^2$ $\downarrow$ $\uparrow$ $2(4x) \rightarrow 4^2$ 2. $x^2 - 12x + \underline{36} = (x - \underline{6})^2$ $\downarrow$ $\uparrow$ $2(6x) \rightarrow 6^2$
3. Solve $x^2 + 2x = 3$ by filling in the blanks and then using the Square Root Property. $$x^2 + 2x + \underline{1} = 3 + \underline{1}$$ Answer: $1, -3$	$x^2 + 2x + \underline{1} = 3 + \underline{1}$ $\downarrow$ $\uparrow$ $\frac{2}{2} = 1^2$ $x^2 + 2x + 1 = 4$ $(x+1)^2 = 4$ $x = -1 \mp 2$ $x+1 = \pm\sqrt{4}$ $x = -1 + 2$ $x+1 = \pm 2$ $x = -1 - 2$ $\boxed{x = 1, -3}$

Example	Notes
4. Solve $x^2+4x-5=0$ by completing the square. $x^2+4x+4=5+4$ $x+2=3$ $x^2+4x+4=9$ $\sqrt{(x+2)^2}=\pm\sqrt{9}$ Answer: 1, −6	$x^2+4x-5=0$ $x^2+4x+4=5+4$ $\frac{4}{2}=2^2=4$ $(x+2)^2=9$ $x+2=\pm\sqrt{9}$ $x=-2\pm3$ $x=-2-3, -2+$ $x=-5,$
5. Solve $x^2+6x+1=0$ by completing the square. $x^2+6x+9=-1+9$ $(x+3)^2 \pm\sqrt{8}$ Answer: $-3+2\sqrt{2}, -3-2\sqrt{2}$	$x^2+6x+9=-1+9$ $3^2\to9$ $(x+3)^2=8$ $x+3=\pm\sqrt{8}$ $x+3=\pm2\sqrt{2}$ $x=-3\pm2\sqrt{2}$ $x=-3+2\sqrt{2}$, $-3-2\sqrt{2}$

Helpful Hints

Recall that when a binomial is squared using FOIL, the coefficient of x is twice the constant of the binomial.

If the equation you are solving contains fractions or if the coefficient of x is odd, then the equation will be more easily solved by using a method other than completing the square.

Concept Check

1. Which method would you use to solve $x^2-8x=-9$? Which method would you use to solve $x^2-8x=-16$?

Practice

Solve by completing the square.

2. $x^2+5x-6=0$

3. $x^2+12x+4=0$

4. $2x^2+16x-96=0$

5. $x^2-4x-16=0$

★ 4:00

Name: __Sydney__   Date: __3/21/16__

Instructor: _____Schnabel_____   Section: _____

Solving Quadratic Equations
Topic 31.5 Solving Quadratic Equations by Using the Quadratic Formula

✓

Vocabulary
standard form   •   Square Root Property   •   quadratic formula

1. For all quadratic equations in the form $ax^2 + bx + c = 0$, you can solve by using the

    __quadratic__ __formula__ , which states that $x = \dfrac{-b \pm \sqrt{b^2 - 4ac}}{2a}$.

Step-by-Step Video Notes
Watch the Step-by-Step Video lesson and complete the examples below.

Example	Notes $\quad X = \dfrac{-b \pm \sqrt{b^2 - 4ac}}{2a}$
1. Solve $3x^2 - x - 2 = 0$ by using the quadratic formula. Identify a, b, and c. Substitute a, b, and c into the quadratic formula. Simplify. $a = 3$ $b = -1$ $C = -2$ $\dfrac{1 \pm \sqrt{25}}{6}$ Answer: $\dfrac{1 \pm 5}{6}$ $\quad x = \dfrac{1+5}{6} = \frac{6}{6} = 1, \dfrac{1-5}{6} = -\frac{4}{6}$	$\dfrac{1 \pm \sqrt{(-1)^2 - 4(3)(-2)}}{2(3)}$ $\dfrac{1 \pm \sqrt{1 - (-24)}}{6}$ $\quad \boxed{x = 1, -\frac{2}{3}}$
2. Solve $x^2 = 6x$ by using the quadratic formula. $x^2 - 6x - 0 = 0$ $1 \quad -6 \quad 0$ $\dfrac{6 \pm \sqrt{36 - 0}}{2}$ $\dfrac{12}{2} \quad \dfrac{0}{2}$ Answer: $6, 0$	$x^2 - 6x = 0$ $a = 1 \ b = -6 \ c = 0$ $\dfrac{-b \pm \sqrt{b^2 - 4ac}}{2a}$ $x = \dfrac{6 \pm \sqrt{(-6)^2 - 4(1)(0)}}{2(1)}$ $\dfrac{6 \pm \sqrt{36 - 0}}{2} = \dfrac{6 \pm \sqrt{36}}{2}$ $\dfrac{6 \pm 6}{2}$ $\quad \boxed{x = 6, 0}$ $\dfrac{6+6}{2} = \dfrac{12}{2} = 6$ $\dfrac{6-6}{2} = \dfrac{0}{2} = 0$

359

Example	Notes
3. Solve $4x^2 + 25 = 20x$ by using the quadratic formula.	$4x^2 - 20x + 25 = 0$ $x = \frac{5}{2}$
	$\dfrac{20 \pm \sqrt{(-20)^2 - 4(4)(+25)}}{2(4)}$ $\quad \frac{20-0}{8}$
	$\dfrac{20 \pm \sqrt{400 - 400}}{8} = \dfrac{20 \pm 0}{8} \quad \dfrac{20}{8} =$
Answer:	
5. Solve $x^2 + 4x - 8 = 0$ by using the quadratic formula.	$x^2 + 4x - 8 = 0 \qquad \dfrac{-4 \pm 4\sqrt{3}}{2}$
	$\dfrac{-4 \pm \sqrt{(4)^2 - 4(1)(-8)}}{2(1)} \quad = \dfrac{-4 \pm \sqrt{48}}{2}$
	$\dfrac{-4 \pm \sqrt{16 - (-32)}}{2}$
Answer: $0, -4$	$\dfrac{-2 - 2\sqrt{3}}{x = -2 + 2\sqrt{3}}$

(handwritten in example 3 box: $4x^2 - 20x + 25 = 0$, a b c, $\dfrac{-b \pm \sqrt{b^2 + 4ac}}{2a}$, $20 \pm \sqrt{400 - 400}$, $\dfrac{20 \pm \sqrt{800}}{2}$, $2\frac{1}{2}, 5$)

(handwritten in example 5 box: $\dfrac{-b \pm \sqrt{b^2 - 4/}}{2a}$, $\dfrac{-4 \pm \sqrt{16 - 8 \cdot 2}}{2}$, $\dfrac{0}{2}$ $\dfrac{-8}{2}$)

Helpful Hints

The quadratic formula is the only method of solving that works for every quadratic equation.

Be sure the equation is in standard form before you identify a, b, and c. If b is positive, then $-b$ in the formula is negative, but if b is negative, then $-b$ is positive in the formula.

Sometimes the value of b or c will be equal to 0. If there is no x term, then the value of b is equal to 0; if there is no constant, then the value of c is equal to 0.

Concept Check

1. How many real solutions does a quadratic equation in the form $ax^2 + bx + c = 0$ have if $b^2 - 4ac = 0$? If $b^2 - 4ac < 0$?

Practice

Solve by using the quadratic formula.

2. $x^2 - 8x + 7 = 0$

4. $9x^2 = 6x - 1$

3. $x^2 + 8x = -2 + 3x$

5. $6x = 4x^2 + 3$

Name: Sydney   Date: 3/21/16
Instructor: Schnabel   Section: _____

Graphing Quadratic Equations
Topic 32.1 Introduction to Graphing Quadratic Equations

Vocabulary
vertex   •   axis of symmetry   •   quadratic equation   ✓

1. The ___vertex___ of a parabola is the lowest or highest point on a parabola.

2. The ___axis of symmetry___ of a parabola is the vertical line through the vertex.

Step-by-Step Video Notes
Watch the Step-by-Step Video lesson and complete the examples below.

Example	Notes
1–3. Determine whether the graph of each quadratic equation opens upward or downward. $y = 2x^2 - 6x + 3$ $y = -5x^2 + 11$ $y = -x^2 + 4x - 1$	1. $y = 2x^2 - 6x + 3$ - upward 2. $y = -5x^2 + 11$ - downward 3. $y = -x^2 + 4x - 1$ - downward

Example	Notes
4–6. If an item is dropped from a height of 400 feet, its height above the ground t seconds after being dropped is given by the equation $h = -16t^2 + 400$. Determine the height of the object for the following values of t. 0 seconds $h = 400$ ft 1 second 384 ft 5 seconds 0 ft	4. $h = -16(0)^2 + 400 = 0 + 400$ $h = 400$ ft. 5. $-16(1)^2 + 400 = -16 + 400$ $h = 384$ ft 6. $-16(5)^2 + 400 = -400 + 400$ $h = 0$ ft

Helpful Hints

Graphs of quadratic equations $y = ax^2 + bx + c$ are parabolas opening upward if $a > 0$ or downward if $a < 0$.

The parabola is symmetric over the axis of symmetry, which means that the graph looks the same on both sides.

Concept Check

1. If the vertex of a parabola is the highest point, does the parabola open upward or downward? What is the sign of the leading coefficient?

Practice

Determine whether the graph of each quadratic equation opens upward or downward.

2. $y = -3x^2 + x + 25$

3. $y = \frac{2}{3}x^2 + 9x - \frac{8}{9}$

Determine the height of an object dropped from a height of 600 feet for each value of t. Use the equation $h = -16t^2 + 600$.

4. $t = 6$ seconds

5. $t = 3$ seconds

Name: Sydney _____   Date: 3/21/16 _____
Instructor: _____ Schnabel _____   Section: _____

Graphing Quadratic Equations
Topic 32.2 Finding the Vertex of a Quadratic Equation

✓

Vocabulary

vertex   •   minimum   •   maximum

1. The ___vertex___ of a parabola is the lowest or highest point on a parabola.

Step-by-Step Video Notes
Watch the Step-by-Step Video lesson and complete the examples below.

Example	Notes
1. Find the vertex of the quadratic equation. Is the vertex a maximum or a minimum? $$y = x^2 - 8x + 15$$ Identify a, b, and c, then find $x = \dfrac{-b}{2a}$. 1 −8 ,5 4 Substitute this x value into the equation and find for the y value of the vertex. $y = (16 - 32 + 15)$ -1 Determine if the vertex is a maximum or minimum. minimum Answer: vertex $(4, -1)$	$y = x^2 - 8x + 15$ $a = 1$ $b = -8$ $c = 15$ $x = \dfrac{8}{2(1)} = \dfrac{8}{2}$ $\boxed{x = 4}$ $y = 4^2 - 8(4) + 15$ $= 16 - 32 + 15$ $\boxed{y = -1}$ * open upward - minimum $\boxed{(4, -1)}$

Example	Notes $-3x^2+12x-6$
2. Find the vertex of the quadratic equation. Is the vertex a maximum or a minimum? $y = 12x - 3x^2 - 6$ Answer: $(2,6)$ maximum	$x = \frac{-12}{2(-3)} = \frac{-12}{-6} = \boxed{2}$ $y = 12(2) - 3(2)^2 - 6$ $24 - 12 - 6$ $\boxed{y = 6}$ $(2,6)$ * maximum
3. Find the vertex of the quadratic equation. Is the vertex a maximum or a minimum? $y = 5x^2 - 7$ Answer: $(0,-7)$ minimum	$y = 5x^2 - 7$ $x = \frac{0}{2(5)} = \frac{0}{10}$ $\boxed{x = 0}$ $y = 5(0)^2 - 7$ $\boxed{y = -7}$ $\boxed{(0,-7)}$ • minimum

Helpful Hints

To find the vertex of a quadratic equation, first identify a, b, and c. Then find the x value of the vertex, which is $x = \dfrac{-b}{2a}$. Substitute this x value into the equation and find the y value of the vertex. If $a > 0$ the vertex is a minimum and if $a < 0$ the vertex is a maximum.

Concept Check

1. Will the vertex of the graph of $y + x^2 = -6x + 4$ be a maximum or a minimum?

Practice

Find the vertex of the quadratic equation. Is the vertex a maximum or a minimum?

2. $y = x^2 - 9x + 8$ 4. $y = -7x^2 + 4x - 3$

3. $y = -4x^2 + 12x - 9$ 5. $y = 3x^2 - 15$

Name: Sydney

Instructor: Schnabel

Date: 3/22/16

Section: _____

Graphing Quadratic Equations
Topic 32.3 Finding the Intercepts of a Quadratic Equation

✓

Vocabulary

intercept • x-intercept • y-intercept • quadratic function

1. To find the __y-intercept__ of a quadratic function, let $x = 0$ and simplify.

2. To find the __x-intercept__ (s) of a quadratic function (if any exist), solve the equation $f(x) = 0$ for x.

Step-by-Step Video Notes

Watch the Step-by-Step Video lesson and complete the examples below.

Example	Notes
1 & 2. Find the y-intercept of each quadratic function. $f(x) = x^2 + 6x + 15$ $(0, 15)$	$f(0) = 0^2 + 6(0) + 15$ $0 + 0 + 15$ $\boxed{(0, 15)}$
$f(x) = 3x^2$ $0 \quad (0, 0)$	$f(0) = 3(0)^2$ $= 0 \quad \boxed{(0,0)}$
3 & 4. Find the x-intercept(s) of each quadratic function, if any exist. $f(x) = x^2 + 2x - 24$ $(6, -6)$ $(0, 4)$ $0 = x^2 + 2x - 24$ $(x+6)(x-4)$	$x =$ $0 = x^2 + 2x - 12 \quad (-6, 0)$ $(x+6)(x-4) = 0 \quad (4, 0)$ $x = -6 \quad x = 4$
$f(x) = x^2 + 1$ $0 = x^2 + 1$ no solutions	$0 = x^2 + 1 \quad \dfrac{0 \pm \sqrt{0^2 - 4(1)(1)}}{2(1)}$ $= \dfrac{\pm \sqrt{-4}}{2} \quad \boxed{\text{no solution}}$

Example	Notes
5. Find the x- and y-intercepts of the quadratic function. $$f(x) = x^2 + 5x - 14$$ *(handwritten)* $x = -7$ $(7, 0)$ $(2, 0)$ $y = -14 \ (0, 14)$ Answer:	*(handwritten)* $0 = x^2 + 5x - 14$ $(x+7)(x-2)$ $x = -7 \quad x = 2$ $\boxed{(-7, 0) \quad (2, 0)}$ $f(0) = (0)^2 + 5(0) - 14$ $= -14$ $\boxed{(0, -14)}$

Helpful Hints

A quadratic function may have zero, one, or two x-intercepts. However, it will always have exactly one y-intercept.

Concept Check

1. Why must a quadratic function have no more than one y-intercept?

Practice

Find the x- and y-intercepts of each quadratic function.

2. $f(x) = x^2 - 6x - 16$

    (handwritten) $(x-8)(x+2)$
    $(8, 0)$
    $(2, 0)$ -16
    $(0, -16)$

4. $f(x) = x^2 + 11$

3. $f(x) = 3x^2 + 18x + 27$

5. $f(x) = x^2 - 9x + 20$

Name: Sydney ___Schnabl___

Date: ___3/22/16___

Instructor: ___Schnabl___

Section: _____

Graphing Quadratic Equations
Topic 32.4 Graphing Quadratic Equations Summary

Vocabulary

intercepts • axis of symmetry • parabola • vertex

1. A parabola is symmetric over the ___axis of symmetry___, which means that the graph looks the same on both sides.

Step-by-Step Video Notes

Watch the Step-by-Step Video lesson and complete the examples below.

Example	Notes
1. Graph the equation $y = x^2$. Find and plot ordered pairs that satisfy the function. Draw a smooth curve through the points. $y = x^2$	$a > 0 =$ upward minimum $a < 0 =$ downward maximum $\begin{array}{c\|c} x & y \\ \hline -2 & 4 \\ -1 & 1 \\ 0 & 0 \\ 2 & 4 \\ 1 & 1 \end{array}$
2. Graph the equation $y = x^2 - 6x + 8$. Determine the direction of the parabola. Find and plot the vertex and the intercepts of the function. Draw a smooth curve through the points. $x = \frac{-b}{2a}$ (3,-1) (1,0)(2,0)	$x^2 - 6x + 8 = 0 \qquad (x-4)(x-2)$ $\frac{6}{2(1)} = \frac{6}{2} = 3 \qquad x = 4 \quad x = 2$ $\begin{array}{c\|c} x & y \\ \hline 4 & 0 \\ 2 & 0 \\ 3 & -1 \\ 0 & 8 \end{array} \qquad x = 3$

*downward

Example	**Notes**

3. Graph the equation $y = -2x^2 + 4x - 3$.

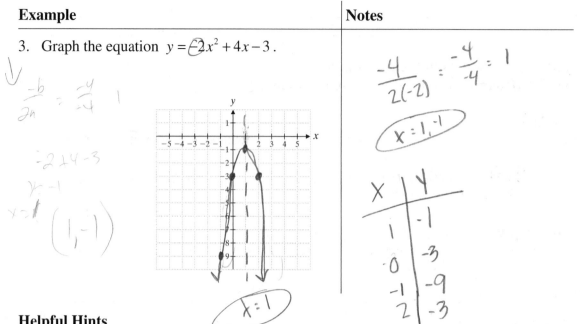

$\frac{-b}{2a} = \frac{-4}{-4} = 1$

$-2 + 4 - 3$

$y = -1$

$x = 1$

$(1, -1)$

$\frac{-4}{2(-2)} = \frac{-4}{-4} = 1$

$x = 1, -1$

X	Y
1	-1
0	-3
-1	-9
2	-3

$x = 1$

Helpful Hints
When graphing a quadratic equation, determine if the parabola opens upward or downward, making the vertex either a minimum or a maximum, respectively. Identify and plot the vertex, the x-intercept(s) and the y-intercept, and other ordered pairs that satisfy the equation, if needed. Draw a smooth curve through the points to form the parabola.

Concept Check
1. For examples 1 through 3, find and graph the axis of symmetry.

Practice
Graph each equation.

2. $y = x^2 - 2x - 3$

3. $y = -2x^2 + 2$

4. $y = -x^2 + 4x - 5$

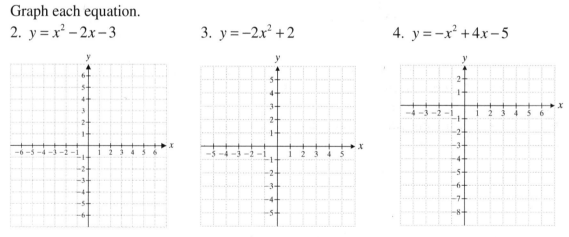

Compound and Quadratic Inequalities
Topic 33.1 Sets

Vocabulary

set • natural number • set intersection • set union • subset • empty set

1. A ___Set___ is a collection of like objects called elements.

2. When all the elements of one set are contained in another set, the contained set is a
___subset___ of the larger set.

Step-by-Step Video Notes

Watch the Step-by-Step Video lesson and complete the examples below.

Example	Notes
1 & 2. List all the elements in each set. Set X is the set of all natural numbers between 4 and 10 . $X = \{5,6,7,8,9\}$ Set Y is the set of all natural numbers between 2 and 7 , inclusive. $Y = \{2,3,4,5,6,7\}$	1. $X = \{5,6,7,8,9\}$ 2. $Y = \{2,3,4,5,6,7\}$
3 & 4. Write in set-builder notation. $A = \{a,e,i,o,u\}$ $A = \{x \mid x$ is a vowel in The alphabet $\}$ Set B is the set of natural numbers between -6 and 0 . No elements $\{\} \varnothing$ or	3. $A = \left\{x \mid x \text{ is a vowel in the alphabet}\right\}$ 4. No elements $\{\} \varnothing$ OR

369
Copyright © 2012 Pearson Education, Inc.

Example	Notes
5. Find the intersection and union of the sets C and D. C = {Odd numbers between 1 and 11, inclusive.} D = {Multiples of 3 between 1 and 15, inclusive.}	$C = \{1, 3, 5, 7, 9, 11\}$ $D = \{3, 6, 9, 12, 15\}$ $C \cap D = \{3, 9\}$ $C \cup D = \{1, 3, 5, 6, 7, 9, 11, 12, 15\}$
6 & 7. Determine if the statement is true or false. Give the reason. $A = \{\text{odd numbers}\}, B = \{\text{integers}\}$, so $A \subseteq B$. $A = \{1, 3, 5, 7...\}$ $B = \{...-2, -1, 0, 1, 2\}$ TRUE $A = \{\text{multiples of 3}\}, B = \{\text{multiples of 4}\}$, so $A \subseteq B$. $\{3, 6, 9, 12...\}$ $\{4, 8, 12, 16...\}$ FALSE	$A \subseteq B$ $\hookrightarrow$ subset

Helpful Hints

The range and domain of functions are often expressed using set-builder notation.

The intersection of two sets is the set of all the elements that are common to both sets. The union of two sets is the set of every element that is in either or both sets.

Concept Check

1. If $A \subseteq B$, what are $A \cap B$ and $A \cup B$?

Practice

Answer each based on the sets $A = \{2, 6, 10, 14...\}, B = \{4, 8, 12, 16...\}$, and $C = \{1, 3, 5, 7,...\}$.

2. Find $A \cup B$.

3. Find $A \cap B$.

4. Is $A \cup B \cup C = \{\text{natural numbers}\}$?

5. Is $\{1, 2, 3, 4, 5\} \subseteq A \cup C$?

Name: _Sydney_   Date: _3/23/16_

Instructor: _Schrabl_   Section: _____

Compound and Quadratic Inequalities
Topic 33.2 Interval Notation

Vocabulary

graph of an inequality   •   infinity   •   negative infinity   •   interval notation

1. A(n) _graph of an inequality_ is a picture that represents all of the solutions of the inequality.

2. The ∞ symbol is used to denote _____. This is the concept of a set without end in the positive direction on a number line.

Step-by-Step Video Notes

Watch the Step-by-Step Video lesson and complete the examples below.

Example	Notes
1 & 2. Graph each inequality and write in interval notation. $x < -2$ $(-\infty, -2)$ $x \geq 0$ $[0, \infty)$	$<$ = less than $\leq$ = less than or equal $>$ = greater $\geq$ = greater than or equal ○ = greater/less than ● = $\leq$ or $\geq$

Example	Notes

Example

3–6. Graph each inequality and write in interval notation.

$-1 \le x < 4$

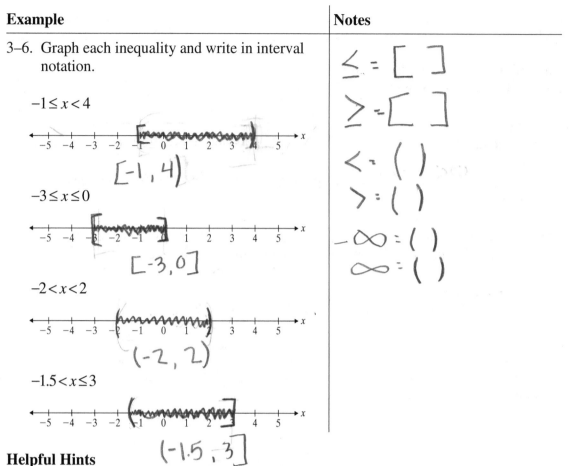

$[-1, 4)$

$-3 \le x \le 0$

$[-3, 0]$

$-2 < x < 2$

$(-2, 2)$

$-1.5 < x \le 3$

$(-1.5, 3]$

Notes

$\le = [\ ]$

$\ge = [\ ]$

$< = (\ )$

$> = (\ )$

$-\infty = (\ )$

$\infty = (\ )$

Helpful Hints

Interval notation is another way to represent the solution to an inequality. It uses two values, the starting point and the ending point of the solution.

In interval notation, a parenthesis is used for endpoints that are not included in the interval and a bracket is used for endpoints that are included in the interval.

Concept Check

1. Describe in words what numbers are included in the interval $[-4.5, \infty)$.

Practice

Graph each inequality and write in interval notation.

2. $x \ge 3$

Write the interval notation for the given inequality.

4.

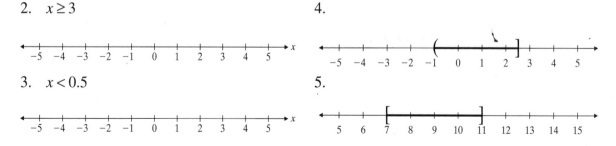

3. $x < 0.5$

5.

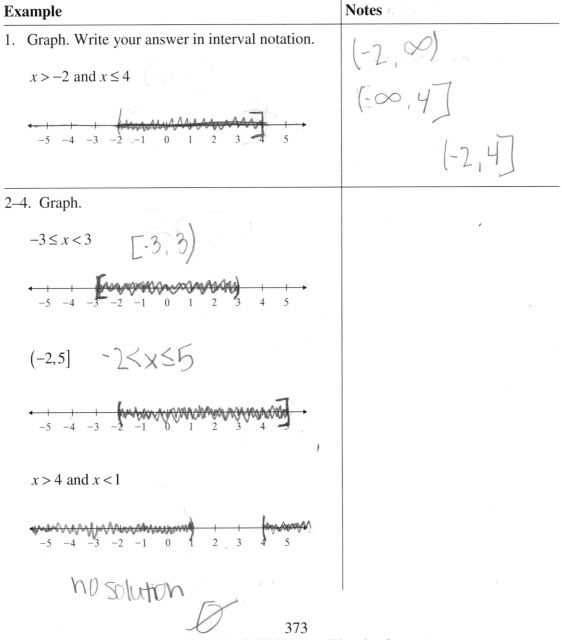

Name: _Sydney Schnabel_   Date: _3/23/16_

Instructor: _Schnabel_   Section: _____

Compound and Quadratic Inequalities
Topic 33.3 Graphing Compound Inequalities

✓

Vocabulary

compound inequality   •   graph of a compound inequality

1. A _compound inequality_ is two or more inequalities separated by "and" or "or."

Step-by-Step Video Notes
Watch the Step-by-Step Video lesson and complete the examples below.

Example	Notes
1. Graph. Write your answer in interval notation. $x > -2$ and $x \leq 4$	$(-2, \infty)$ $[-\infty, 4]$ $[-2, 4]$

2–4. Graph.

$-3 \leq x < 3$ $[-3, 3)$

$(-2, 5]$ $-2 < x \leq 5$

$x > 4$ and $x < 1$

no solution ∅

373
Copyright © 2012 Pearson Education, Inc.

Example	Notes

6 & 7. Graph. Write your answer in interval notation.

$x > 3$ or $x \leq -2$ $\quad (-\infty, -2] \cup (3, \infty)$

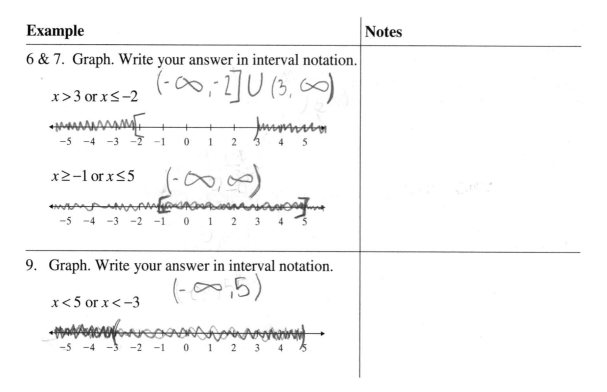

$x \geq -1$ or $x \leq 5$ $\quad (-\infty, \infty)$

9. Graph. Write your answer in interval notation.

$x < 5$ or $x < -3$ $\quad (-\infty, 5)$

Helpful Hints

The solution to a compound inequality using the word "and" is the set of all the points on a number line which satisfies both the inequalities. The interval of the graph is the intersection of the intervals of each part of the inequality.

The solution to a compound inequality using the word "or" is the set of all the numbers on a number line which satisfies either one of the inequalities. The interval of the graph is the union of the intervals of each part of the inequality.

Concept Check

1. Can you give examples of any numbers not in the solution set of $x \leq 0$ or $x > -5$?

Practice

Graph. Write your answer in interval notation.

2. $y \geq -3$ and $y < 5$

4. $x < 2$ or $x > 2$

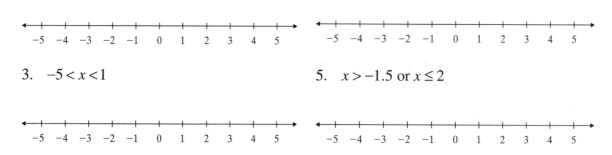

3. $-5 < x < 1$

5. $x > -1.5$ or $x \leq 2$

Name: Sydney

Instructor: Schnabel

Date: 3/23/16

Section: _____

Compound and Quadratic Inequalities
Topic 33.4 Solving Compound Inequalities

Vocabulary

compound inequality • interval notation • inequality

1. To solve a(n) _Compound inequality_, solve each of the inequalities for the variable and express the solution in _interval notation_.

Step-by-Step Video Notes
Watch the Step-by-Step Video lesson and complete the examples below.

Example	Notes
1. Solve and graph $x+1>-2$ and $x-2\le 2$. Then write the solution in interval notation.	$x+1>-2$ $x-2\le 2$ $x>-3$ and $x\le 4$ $(-3,4]$
2. Solve $3x+2\ge 14$ or $2x-1\le -7$ for x and graph the solution. Then write the solution using interval notation.	$3x+2\ge 14$ $2x-1\le -7$ $3x\ge 12$ $2x\le -6$ $x\ge 4$ or $x\le -3$ $(-\infty,-3]\cup[4,\infty)$

Example	Notes
4. Solve $6+x<9$ or $5x+2>-18$ for x and graph the solution. Then write the solution using interval notation.	$6+x<9$ OR $5x+2>-18$ $x<3$ OR $5x>-20$ $x>-4$ $(-\infty, \infty)$
5. Solve $-9 \leq 3a-3 < 9$ for a and graph the solution. Then write the solution using interval notation.	$-9 \leq 3a-3 < 9$ $-9 \leq 3a-3$ and $3a-3<9$ $-6 \leq 3a$ $3a<12$ $-2 \leq a$ and $a<4$ $(-2, 4]$

Helpful Hints

To solve an inequality, use the same procedure used to solve equations, except reverse the direction of the inequality if you multiply or divide by a negative number.

Concept Check

1. Write the two separate inequalities you must solve to find the solution to the compound inequality $-2 \leq -4x+6 < 14$.

Practice

Solve for x and graph the solution. Then write the solution using interval notation.

2. $x+2 \geq -1$ and $x+7 \leq 9$ 4. $-7x+8<-27$ and $3x+10 \leq 4$

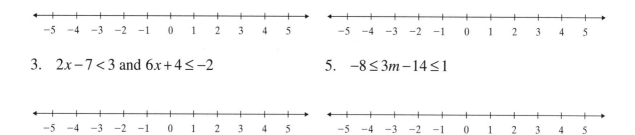

3. $2x-7<3$ and $6x+4 \leq -2$ 5. $-8 \leq 3m-14 \leq 1$

Compound and Quadratic Inequalities
Topic 33.5 Solving Quadratic Inequalities

Vocabulary
quadratic inequality   •   compound inequality   •   quadratic formula

1. To solve a _____, replace the inequality symbol with an equal sign. Solve the resulting equation to find the solutions, or boundary numbers. Use these numbers to separate the number line into regions to test the original for solutions.

Step-by-Step Video Notes
Watch the Step-by-Step Video lesson and complete the examples below.

Example	Notes
1. Solve $(x-3)(x+1)>0$ and graph the solution. Write the solution in interval notation.	

Example	Notes
2. Solve $2x^2+x-6\le 0$ and graph the solution. Write the solution in interval notation.	

Example	Notes
3. Solve $x^2 - 4x + 4 > 0$ and graph the solution. Write the solution in interval notation. -5 -4 -3 -2 -1 0 1 2 3 4 5	
4. Solve $x^2 + 4x - 6 \geq 0$ and graph the solution. Write the solution in interval notation. -5 -4 -3 -2 -1 0 1 2 3 4 5	

Helpful Hints

When testing to see which regions of the number line satisfy the original inequality, choose numbers that make the calculations easier. Integers are generally easier to use in calculations than decimals or fractions.

Concept Check

1. Is there a quadratic inequality for which the solution would be all real numbers? No solution?

Practice

Solve and graph. Then write the solution in interval notation.

2. $x^2 - 2x - 15 > 0$ 4. $x^2 + 3x - 9 \leq 0$

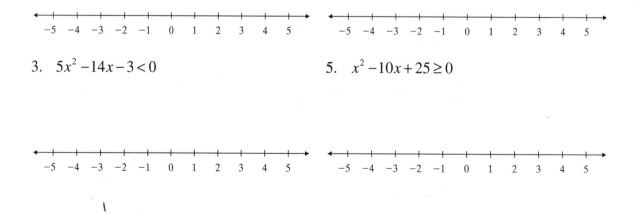

3. $5x^2 - 14x - 3 < 0$ 5. $x^2 - 10x + 25 \geq 0$

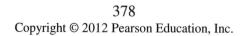

Name: Sydney

Instructor: Schnabel

Date: 3/24/16

Section: _____

✓

Absolute Value Equations and Inequalities
Topic 34.1 Introduction to Absolute Value Equations

Vocabulary

absolute value   •   absolute value equation rule

1. The __absolute value equation rule__ states that if $|x| = a$, and a is a positive real number, then $x = a$ or $x = -a$.

Step-by-Step Video Notes

Watch the Step-by-Step Video lesson and complete the examples below.

Example	Notes
2–4. Solve. $\quad$ $|x| = 8.35$ $\quad$ $|y| = 0$ $\quad$ $|x| = -11$	2. $|x| = 8.35$ $\quad$ $\boxed{x = 8.35}$ $\boxed{x = -8.35}$ 3. $|y| = 0$ $\quad$ $\boxed{y = 0}$ 4. $|x| = -11$ $\boxed{\text{no solution}}$
5. Solve $|x| + 1 = 10$. Isolate the absolute value expression. Write as two equations according to the absolute value equation rule. Solve by applying the rule for absolute value equations. Answer: $\boxed{x = 9, -9}$	$|x| + 1 = 10$ $|x| = 9$ $x = 9 \qquad x = -9$ $9 + 1 = 10 \qquad\qquad |-9| + 1 = 10$ $9 + 1 = 10 \qquad\qquad 9 + 1 = 0$ $10 = 10 ✓ \qquad\quad 10 = 10 ✓$

379

Example	Notes
6. Solve $-6|x| = -72$.	$-6|x| = -72$ $x = 12, -12$
	$|x| = 12$ $-6(-12) = -72$
	$-72 = -72$ ✓
Answer: $x = 12, \quad x = -12$	$-6(12) = -72$
	$-72 = -72$ ✓

7 & 8. Write an equation using absolute value to represent the given graph.

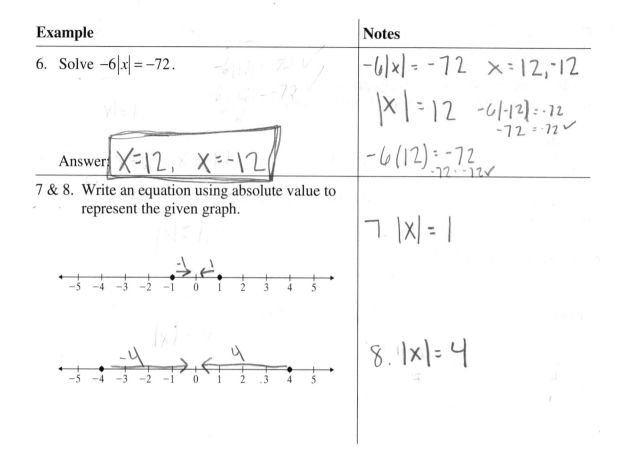

$7. \ |x| = 1$

$8. \ |x| = 4$

Helpful Hints
The absolute value of a number is the distance between that number and zero on a number line.

Concept Check
1. Why does the equation $|x| = -6$ have no solution?

Practice
Solve.

2. $|n| = 6\dfrac{7}{8}$

3. $|x| + 5.8 = 3.2$

4. $\dfrac{1}{3}|a| = 6$

5. $-4|x| = -5.2$

Name: Sydney

Instructor: Schnabel

Date: 3/24/16

Section: _____

Absolute Value Equations and Inequalities
Topic 34.2 Solving Basic Absolute Value Equations

Vocabulary

absolute values • absolute value equations rule

1. The first step to solving an absolute value equation is to isolate the _absolute values_

Step-by-Step Video Notes

Watch the Step-by-Step Video lesson and complete the examples below.

Example	Notes
1. Solve $|x+1|=6$. Answer:	$|x+1|=6$ $x+1=6 \qquad x+1=-6$ $\boxed{x=5} \qquad \boxed{x=-7}$ $\qquad\qquad\qquad |-7+1|=6$ $5+1=6 \quad 6=6 ✓ \qquad |-6|=6$ $\qquad\qquad\qquad\qquad 6=6 ✓$
2. Solve $|5y|=35$. Answer:	$5y=35 \qquad 5y=-35$ $\boxed{y=7} \qquad \boxed{y=-7}$ $|5(7)|=35 \qquad |5(-7)|=35$ $35=35 ✓ \qquad |-35|=35$ $\qquad\qquad\qquad 35=35 ✓$
3. Solve $\left|\dfrac{1}{2}x-1\right|=5$. Write the absolute value equation as two separate equations. Answer:	$\frac{1}{2}x-1=5 \qquad \frac{1}{2}x-1=-5$ $2\cdot\frac{1}{2}x=6\cdot2 \qquad 2\cdot\frac{1}{2}x=-4\cdot2$ $\boxed{x=12} \qquad \boxed{x=-8}$ $\left|\frac{1}{2}(12)-1\right|=5 \qquad \left|\frac{1}{2}(8)-1\right|=-5$ $|6-1|=5 \qquad |-4-1|=5$ $5=5 ✓ \qquad |-5|=5$ $\qquad\qquad\qquad 5=5 ✓$

Example	Notes

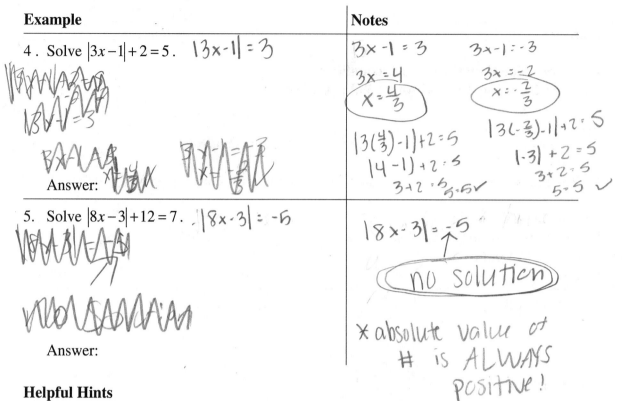

Example

4. Solve $|3x-1|+2=5$. $|3x-1|=3$

Answer:

Notes

$3x-1=3 \qquad 3x-1=-3$

$3x=4 \qquad\qquad 3x=-2$

$\boxed{x=\dfrac{4}{3}} \qquad \boxed{x=-\dfrac{2}{3}}$

$|3(\tfrac{4}{3})-1|+2=5 \qquad |3(-\tfrac{2}{3})-1|+2=5$

$|4-1|+2=5 \qquad\qquad |-3|+2=5$

$3+2=5 \quad 5=5\checkmark \qquad\qquad 3+2=5$

$5=5\checkmark$

5. Solve $|8x-3|+12=7$. $|8x-3|=-5$

Answer:

$|8x-3|=-5$

$\boxed{no\ solution}$

$\ast$ absolute value of # is ALWAYS positive!

Helpful Hints

Remember, the absolute value of a number is the distance between that number and zero on a number line.

If $|x|=a$, and a is a positive real number, then $x=a$ or $x=-a$.

Concept Check

1. If $|7x-3y|=4z$, then $7x-3y$ is equal to what two numbers?

Practice

Solve.

2. $|x+7|=13$

4. $\left|\dfrac{2}{3}-\dfrac{1}{6}x\right|=3$

3. $|9x-6|=12$

5. $|2x-1|-5=4$

Absolute Value Equations and Inequalities
Topic 34.3 Solving Multiple Absolute Value Equations

Vocabulary
absolute value expressions   •   multiple absolute value equations

1. In order for two _____ to be equal, the expressions inside the absolute value bars must be either equal to each other or opposites of each other.

Step-by-Step Video Notes
Watch the Step-by-Step Video lesson and complete the examples below.

Example	Notes
1. Solve $\lvert 3x+4 \rvert = \lvert x \rvert$. Write the absolute value equation as two separate equations. $3x+4 = x$ $\quad$ $3x+4 = -x$ $-x \;-4 \;-x$ $\quad$ $+x \;-4 \;+x$ Solve the resulting equations. $2x = -4$ $\qquad$ $4y = -4$ $\quad x = -2$ $\qquad$ $x = -1$ Answer: $x = -2, -1$	
2. Solve $\lvert 3x-4 \rvert = \lvert x+6 \rvert$. Write the absolute value equation as two separate equations. $3x-4 = -x-6$ $\quad$ $3x-4 = x+6$ $+x \qquad +x$ $\qquad$ $-x \qquad -x$ $4y = -2$ $\qquad\quad$ $2x = 10$ $\quad x = -\frac{1}{2}$ $\qquad\quad$ $x = 5$ Answer:	

Example	**Notes**
3. Solve $|x+3| = |x-5|$.	

$$X+3 = -X+5 \qquad X+3 = X-5$$
$${-X} \qquad {-X}$$
$$2X = 2 \qquad 0 = -8$$
$$X = 1$$

Answer: $X = 1$

4. Solve $|2x-4| = |4-2x|$.

$$2X-4 = -4+2X \qquad 2X-4 = 4-2X$$
$${-2X} \qquad 4X = 8$$
$$0 = 0 \qquad X = 2$$

Answer: $X = 2$ $\;$ (Infinite Solutions)

Helpful Hints

When solving equations with an absolute value expression alone on both sides of the equation, if the expressions inside the absolute value bars are equivalent or opposites, the equation is an identity, and all real numbers are solutions of the equation.

Concept Check

1. Marissa reasons that since $|x| = |y|$, then $x = y$. Is she correct? Explain.

Practice

Solve.

2. $|4x+12| = |2x|$

4. $\left|4-x\right| = \left|\dfrac{x}{3}+2\right|$

3. $|x-4| = |4x+11|$

5. $|3x+2| = |3x-8|$

Name: _____   Date: _____
Instructor: _____   Section: _____

Absolute Value Equations and Inequalities
Topic 34.4 Solving Absolute Value Inequalities

Vocabulary

$|x| < a$ • $|x| > a$ •   absolute value equation rule

1. If a is a positive real number, and _____, then $-a < x < a$ or $x > -a$ and $x < a$.

Step-by-Step Video Notes
Watch the Step-by-Step Video lesson and complete the examples below.

Example	Notes
1 & 2. Graph. Write the answer in interval notation.	

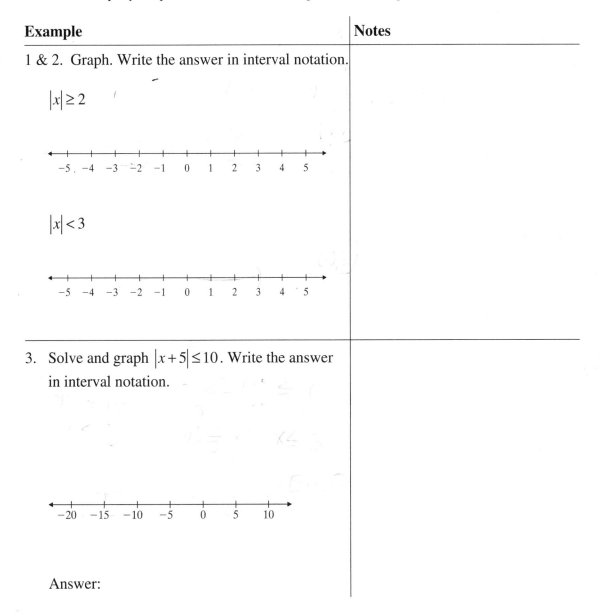

$|x| \geq 2$

$|x| < 3$

3. Solve and graph $|x + 5| \leq 10$. Write the answer in interval notation.

Answer:

Example	Notes

4 . Solve and graph $|-2x-1| \geq 7$. Write the answer in interval notation.

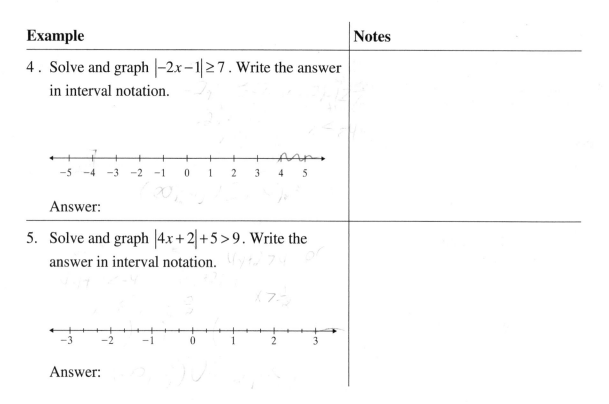

Answer:

5. Solve and graph $|4x+2|+5>9$. Write the answer in interval notation.

Answer:

Helpful Hints

If $|x|>a$ and a is a positive real number, then $x<-a$ or $x>a$. The interval notation for the solution is $(-\infty,a)\cup(a,\infty)$.

When graphing inequalities, use an open circle (or parenthesis) for < or > and a closed circle (or bracket) for ≤ or ≥.

Concept Check

1. Sketch the number line graph that displays the interval $(-\infty,-3]\cup[6,\infty)$.

Practice

Solve and graph. Write the answer in interval notation.

2. $|x+5|\leq 4$ 4. $|-4x+3|\geq 11$

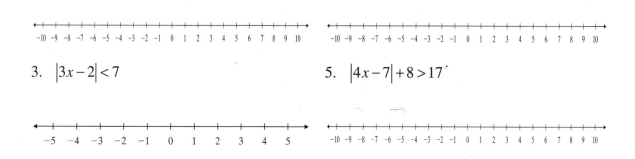

3. $|3x-2|<7$ 5. $|4x-7|+8>17$

Name: _____   Date: _____

Instructor: _____   Section: _____

Conic Sections
Topic A.1 The Circle

Vocabulary
conic section   •   circle   •   standard form

1. A _____ is a shape that can be formed by slicing a cone with a plane.

2. In a given plane, a _____ is a set of all points that are a fixed distance from a center point.

Step-by-Step Video Notes
Watch the Step-by-Step Video lesson and complete the examples below.

Example	Notes
2. Find the center and radius of the following circle. Then sketch its graph. $$(x+2)^2 + (y-3)^2 = 25$$ 	
3. Write the equation of the circle in standard form. Find the radius and center of the circle and sketch its graph. $$x^2 + 2x + y^2 + 6y + 6 = 0$$ 	

Example	Notes
5. Write the equation of the circle in standard form with the given center and radius. Center $(8,-2)$; $r = 7$ Answer:	
7. A Ferris wheel has a radius, r, of 25.3 feet. The height of the tower, t, is 31.8 feet. The distance, d, from the origin to the base of the tower is 44.8 feet. Find the standard form of the equation of the circle represented by the Ferris wheel. Answer:	

Helpful Hints

The standard form of the equation of a circle with center at (h,k) and radius r is $(x-h)^2 + (y-k)^2 = r^2$.

Be careful of the signs when making calculations with h and k. For example, the equation $(x+2)^2 + (y-3)^2 = 25$ has a center of $(-2,3)$, not $(2,3)$.

Concept Check
1. Explain why a circle can have negative coordinates for its center but cannot have a negative radius.

Practice
For the circle $x^2 - 2x + y^2 + 2y = 2$, determine the following information.

2. Write the equation of the circle in standard form.

3. Identify the center and radius of the circle.

4. Graph the circle and label the center and radius.

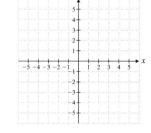

Name: _____   Date: _____
Instructor: _____   Section: _____

Conic Sections
Topic A.2 The Parabola

Vocabulary
conic section   •   parabola   •   standard form

1. A _____ is the graph of a quadratic function.

Step-by-Step Video Notes
Watch the Step-by-Step Video lesson and complete the examples below.

Example	**Notes**
1. Graph the parabola $y = 2(x-1)^2 - 3$. Determine if the parabola is vertical or horizontal. Identify a, h, and k. $a = \boxed{}$, $h = \boxed{}$, and $k = \boxed{}$ Determine the direction the parabola opens. $a \boxed{}\, 0$; the parabola opens _____. Find the vertex. $(\boxed{}, \boxed{})$	

Example	Notes
2. Graph the parabola $x = (y+2)^2 - 3$. 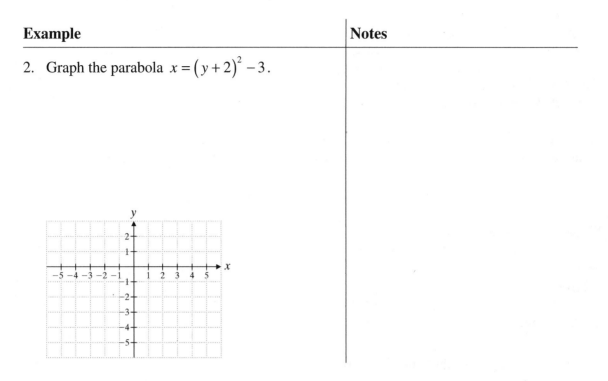	

Helpful Hints

The graph of $y = a(x-h)^2 + k$, where $a \neq 0$, is a vertical parabola. The parabola opens upward if $a > 0$ and downward if $a < 0$. The vertex of the parabola is (h,k). The axis of symmetry is the line $x = h$.

The graph of $x = a(y-k)^2 + h$, where $a \neq 0$, is a horizontal parabola. The parabola opens to the right if $a > 0$ and opens to the left if $a < 0$. The vertex of the parabola is (h,k). The axis of symmetry is the line $y = k$.

Concept Check

1. Do parabolas always have both x and y-intercepts? Explain.

Practice

Graph the parabolas. Label the vertices and axes of symmetry.

2. $y = (x-2)^2 + 3$ 3. $y = -(x+3)^2 - 1$ 4. $x = -3(y+3)^2 - 2$

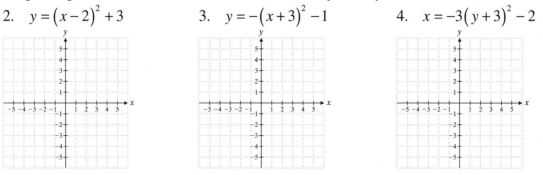

Conic Sections
Topic A.3 The Ellipse

Vocabulary

conic section   •   ellipse   •   focus

1. A(n) _____ is the set of points in a plane such that for each point in the set, the sum of its distances to two fixed point is constant.

Step-by-Step Video Notes

Watch the Step-by-Step Video lesson and complete the examples below.

Example	Notes
2. Graph the ellipse. Label the intercepts. $x^2 + 3y^2 = 12$ 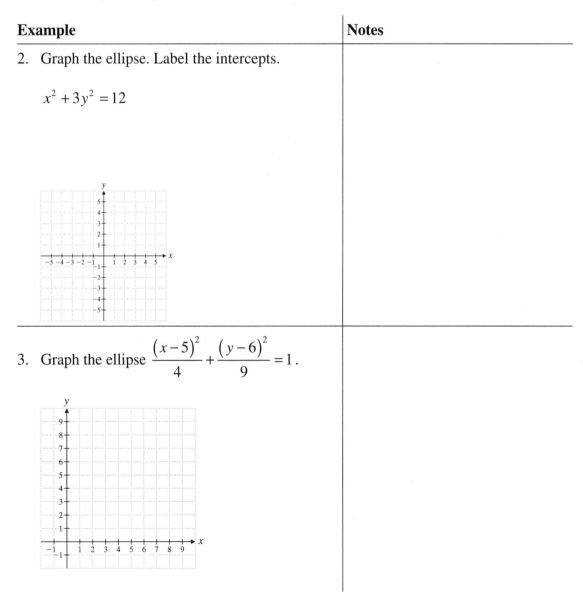	
3. Graph the ellipse $\dfrac{(x-5)^2}{4} + \dfrac{(y-6)^2}{9} = 1$.	

Example	Notes

4. The orbit of Venus is an ellipse with the Sun as a focus. If we say that the center of the ellipse is at the origin, an approximate equation for the orbit is $\dfrac{x^2}{5013} + \dfrac{y^2}{4970} = 1$, where x and y are measured in millions of miles. Find the largest possible distance across the ellipse. Round your answer to the nearest million miles.

Helpful Hints

An ellipse centered at the origin has the equation $\dfrac{x^2}{a^2} + \dfrac{y^2}{b^2} = 1$, where $a > 0$ and $b > 0$. The vertices of this ellipse are at $(a, 0)$, $(-a, 0)$, $(b, 0)$, and $(-b, 0)$.

The standard form of the equation of an ellipse is $\dfrac{(x-h)^2}{a^2} + \dfrac{(y-k)^2}{b^2} = 1$, where $a > 0$ and $b > 0$.

Concept Check
1. Explain why every circle is an ellipse but not every ellipse is a circle.

Practice
Graph the ellipses. Label the vertices.

2. $\dfrac{x^2}{16} + \dfrac{y^2}{25} = 1$

3. $\dfrac{(x+1)^2}{9} + \dfrac{(y-2)^2}{4} = 1$

4. $x^2 + \dfrac{(y+3)^2}{4} = 1$

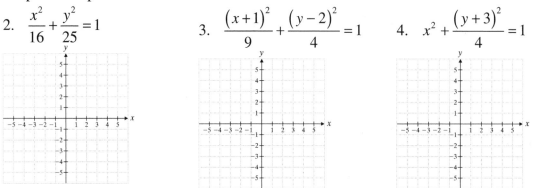

392

Name: _____   Date: _____

Instructor: _____   Section: _____

Conic Sections
Topic A.4 The Hyperbola

Vocabulary
conic section   •   hyperbola   •   asymptotes of hyperbolas
fundamental rectangle of a hyperbola

1. A(n) _____ is the set of points in a plane such that for each point in the set, the absolute value of the difference of its distances to two fixed points (called foci) is a constant.

Step-by-Step Video Notes
Watch the Step-by-Step Video lesson and complete the examples below.

Example	Notes
1. Graph the hyperbola $\dfrac{x^2}{25} - \dfrac{y^2}{16} = 1$.	
2. Graph the hyperbola $4y^2 - 7x^2 = 28$.	

Example	Notes

3. Graph the hyperbola $\dfrac{(x-4)^2}{9} - \dfrac{(y-5)^2}{4} = 1$.

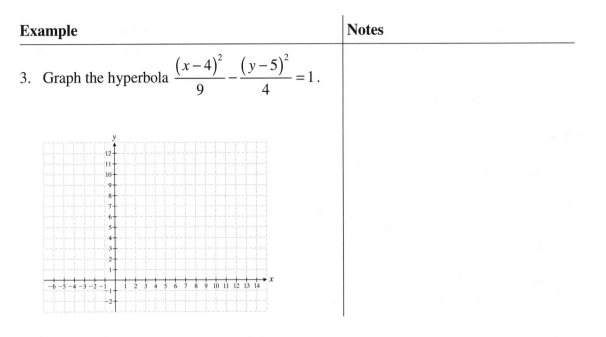

Helpful Hints

Let a and b be any positive real numbers. A horizontal hyperbola with center at (h,k) and vertices $(h-a,k)$ and $(h+a,k)$ has the equation $\dfrac{(x-h)^2}{a^2} - \dfrac{(y-k)^2}{b^2} = 1$.

Let a and b be any positive real numbers. A vertical hyperbola with center at (h,k) and vertices $(h,k-b)$ and $(h,k+b)$ has the equation $\dfrac{(y-k)^2}{b^2} - \dfrac{(x-h)^2}{a^2} = 1$.

Concept Check
1. Compare and contrast the four types of conic sections (circles, parabolas, ellipses, and hyperbolas).

Practice
Graph the hyperbolas.

2. $9x^2 - 5y^2 = 45$

3. $-\dfrac{(x+1)^2}{9} + \dfrac{(y-2)^2}{4} = 1$

4. $\dfrac{(x-1)^2}{4} - y^2 = 1$

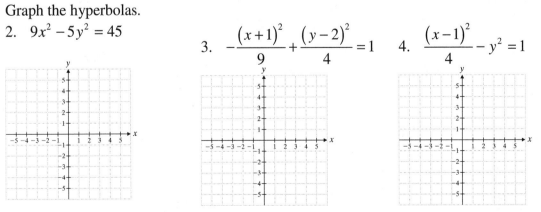

Logarithmic and Exponential Functions
Topic B.1 Evaluating Exponential and Logarithmic Expressions

Vocabulary
logarithm   •   exponential expression

1. The _____, base b, of a positive number, x, is the power (exponent) to which the base b must be raised to produce x and is represented by the expression $\log_b x$.

Step-by-Step Video Notes
Watch the Step-by-Step Video lesson and complete the examples below.

Example	Notes

1–3. Evaluate each exponential expression for the given value of x.
 	
  5^x; $x = 2$
 	
  4^x; $x = 3$
 	
  2^x; $x = 5$	

5 & 6. Evaluate the exponential expression for the given value of x.
 	
  2^x; $x = -1$
 	
  6^x; $x = -2$	

Example	Notes
7–10. Evaluate each logarithmic expression. $\log_2 4$ $\log_5 5$ $\log_3 \dfrac{1}{27}$ $\log_4 1$	

Helpful Hints

No matter the base b, $\log_b x$ for $0 < x < 1$ is always negative, whereas $\log_b x$ is always positive for $1 < x$.

No matter the base b, $\log_b 1 = 0$.

No matter the base b, $\log_b b = 1$.

Concept Check

1. If $4^{10} = 1,048,576$, what is $\log_4 1,048,576$?

Practice

Evaluate the exponential expression for the given values of x.

2. 9^x; $x = 3$, $x = 0$

3. 3^x; $x = -2$, $x = -4$

Evaluate each logarithmic expression.

4. $\log_4 64$

5. $\log_5 \dfrac{1}{25}$

6. $\log_{12,589,654} 1$

Logarithmic and Exponential Functions
Topic B.2 Graphing Exponential Functions

Vocabulary
exponential function　　•　　e　　•　　base

1. A(n) _____ is a function of the form $f(x) = b^x$, where $b > 0$, $b \neq 1$, and x is a real number.

2. The number _____ is an irrational number that occurs in many formulas that describe real-world phenomena, such as the growth of cells and radioactive decay.

Step-by-Step Video Notes
Watch the Step-by-Step Video lesson and complete the examples below.

Example	Notes
1. Graph the exponential function $f(x) = 2^x$.	
2. Graph the exponential function $f(x) = \left(\dfrac{1}{2}\right)^x$.	

Example	Notes

3. Graph the exponential function $f(x) = e^x$.

Helpful Hints

An asymptote is a line that the graph of a function gets closer to as the value of x gets larger or smaller. The x-axis is an asymptote for every exponential function of the form $f(x) = b^x$.

The domain of an exponential function $f(x) = b^x$ is the set of all real numbers, whereas the range of an exponential function $f(x) = b^x$ is the set of all positive real numbers.

Concept Check

1. Explain why the graph of $f(x) = b^x$ always passes through $(0,1)$ and $(1,b)$.

Practice

Graph the exponential functions.

2. $f(x) = 3^x$

3. $f(x) = \left(\dfrac{1}{4}\right)^x$

4. $f(x) = \left(\dfrac{1}{e}\right)^x$

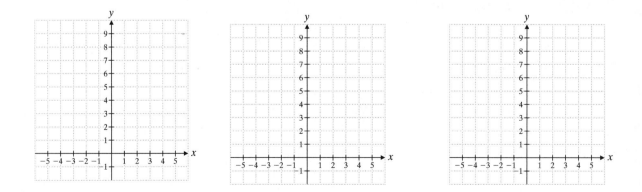

Logarithmic and Exponential Functions
Topic B.3 Solving Simple Exponential Equations and Applications

Vocabulary
property of exponential equations   •   compound interest formula
variable compound interest formula   •   exponential decay formula

1. The _____ states that if $b^x = b^y$, then $x = y$ for $b > 0$ and $b \neq 1$.

Step-by-Step Video Notes
Watch the Step-by-Step Video lesson and complete the examples below.

Example	Notes
1 & 2. Solve the exponential equations. $2^x = 4$ $3^x = 1$	
3 & 4. Solve the exponential equations. $2^x = \dfrac{1}{16}$ $8^{3x-1} = 64$	

Example	Notes
6. If we invest \$8000 in a fund that pays 15% annual interest compounded monthly, how much will we have in the account after 6 years?	
7. The decay constant for americium 241 is $k = -0.0016008$. If 10 milligrams (mg) of americium 241 is sealed in a laboratory container today, how much will still be present in 50 years?	

Helpful Hints

To solve an exponential equation, get the same base on both sides of the equation, if possible. Then, set the exponents on each side of the equation equal to each other. If necessary, solve this equation.

If a principle amount, P, is invested at an interest rate, r, that is compounded n times a year, the amount of money, A, accumulated after t years is $A = P\left(1 + \dfrac{r}{n}\right)^{nt}$.

Concept Check

1. Explain how the compound interest formula is a special case of the variable compound interest formula.

Practice

Solve the exponential equations.

2. $5^x = 125$

3. $2^{2x-1} = \dfrac{1}{8}$

4. If a young married couple invests \$9000 in a mutual fund that pays 10% interest compounded annually, how much will they have in 5 years?

Logarithmic and Exponential Functions
Topic B.4 Converting Between Exponential and Logarithmic Forms

Vocabulary

logarithmic equation   •   exponential equation   •   logarithm

1. The _____ $y = \log_b x$ is the same as the _____ $x = b^y$, where $b > 0$ and $b \neq 1$.

Step-by-Step Video Notes
Watch the Step-by-Step Video lesson and complete the examples below.

Example	Notes
1. Write $81 = 3^4$ in logarithmic form.	
2–4. Write in logarithmic form. $49 = 7^2$ $512 = 8^3$ $\dfrac{1}{100} = 10^{-2}$	

Example	Notes
5. Write $2 = \log_5 25$ in exponential form.	
6–8. Write in exponential form. $\dfrac{1}{2} = \log_{16} 4$ $0 = \log_{17} 1$ $-4 = \log_{10}\left(\dfrac{1}{10,000}\right)$	

Helpful Hints

The logarithm, base b, of a positive number, x, is the power (exponent) to which the base b must be raised to produce x and is represented by the expression $\log_b x$.

Be careful with signs when changing forms.

Concept Check

1. When converting $14^2 = 196$ to logarithmic form, Melissa wrote $196 = \log_2 14$. Explain her error.

Practice

Write in logarithmic form.

2. $13^2 = 169$

3. $\left(\dfrac{1}{7}\right)^4 = \dfrac{1}{2401}$

Write in exponential form.

4. $1 = \log_{4289} 4289$

5. $-3 = \log_6 \dfrac{1}{216}$

Logarithmic and Exponential Functions
Topic B.5 Solving Simple Logarithmic Equations

Vocabulary
logarithmic equation   •   exponential equation   •   logarithm

1. To solve a(n) _____, convert it to a(n) _____ and solve.

Step-by-Step Video Notes
Watch the Step-by-Step Video lesson and complete the examples below.

Example	Notes
1. Solve the logarithmic equation $\log_2 x = 4$. The exponential equation is $\square^{\square} = x$. $x = \square$ Check the solution. Answer:	
2–4. Solve the logarithmic equations. $3 = \log_3 x$ $\log_{10} x = -3$ $\log_{25} x = \dfrac{1}{2}$	

Example	Notes
5 & 6. Solve the logarithmic equations. $\log_7 343 = y$ $\log_8 \left(\dfrac{1}{64} \right) = y$	

7 & 8. Solve the logarithmic equations. $2 = \log_b 121$ $\log_b 81 = 4$	

Helpful Hints

To solve an exponential equation, remember to first get the same bases on both sides of the equation. Then set the exponents equal to each other.

To solve an exponential equation for the base, try to get the exponents on both sides of the equation equal, and then set the bases equal to each other.

Concept Check

1. While solving the equation $\log_b 4 = 2$, Robert got the answer $b = 16$. Explain his error.

Practice

Solve the logarithmic equations.

2. $\log_3 x = 4$

3. $\log_8 x = \dfrac{1}{3}$

4. $\log_b 128 = 7$

5. $\log_5 \left(\dfrac{1}{125} \right) = y$

Logarithmic and Exponential Functions
Topic B.6 Graphing Logarithmic Functions

Vocabulary
logarithmic function   •   exponential function   •   asymptote

1. A(n) _____ is a function of the form $f(x) = \log_b x$, where $b > 0$, $b \neq 1$, and x is a positive real number.

Step-by-Step Video Notes
Watch the Step-by-Step Video lesson and complete the examples below.

Example	Notes
1. Graph the logarithmic function $f(x) = \log_2 x$.	
2. Graph the logarithmic function $f(x) = \log_{1/3} x$.	

Example	Notes
3. Graph $f(x) = \log_2 x$ and $f(x) = 2^x$ on the same set of axes. 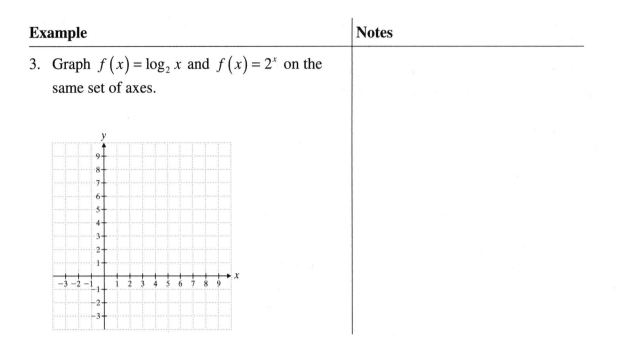	

Helpful Hints

The y-axis is an asymptote for every logarithmic function of the form $f(x) = \log_b x$.

The domain of an logarithmic function $f(x) = \log_b x$ is the set of all positive real numbers, whereas the range of an logarithmic function $f(x) = \log_b x$ is the set of all real numbers.

Concept Check

1. Explain how the domain and range of an exponential function differ from those of a logarithmic function.

Practice
Graph the logarithmic functions.

2. $f(x) = \log_{1/2} x$ 3. $f(x) = \log_e x$ 4. $f(x) = \log_3 x$

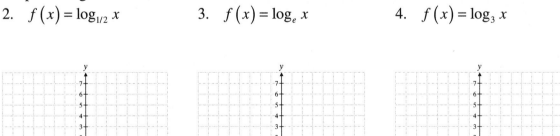

Solving Logarithmic and Exponential Equations
Topic C.1 Properties of Logarithms

Vocabulary
logarithm of a quotient • logarithm of a product • logarithm of a number raised to a power

1. The _____ property states that for any positive real numbers, M and N, and any positive base, $b \neq 1$, $\log_b MN = \log_b M + \log_b N$.

Step-by-Step Video Notes
Watch the Step-by-Step Video lesson and complete the examples below.

Example	Notes
1 & 2. Write each logarithm as a sum of logarithms. $\log_3 AB$ $\log_5 (7 \cdot 11)$	
4. Write the logarithm as a difference of logarithms. $\log_3 \left(\dfrac{29}{7} \right)$ Answer:	
5. Write the expression as a single logarithm. $\log_b 45 - \log_b 9$ Answer:	

Example	Notes
6 & 7. Express each logarithm as a product. $\log_5 b^{10}$ $\log_8 \sqrt{w}$	
8. Write the expression as a single logarithm. $\frac{1}{3}\log_b x + 2\log_b w - 3\log_b z$ Answer:	

Helpful Hints

Note that $\left(\log_b M\right)\left(\log_b N\right) \neq \log_b M + \log_b N$ and $\dfrac{\log_b M}{\log_b N} \neq \log_b M - \log_b N$. These are the most common mistakes made when using the properties of logarithms.

For all the properties of logarithms, the base of the logarithm(s) must be the same for the property to apply.

Concept Check

1. Using the properties of exponents, explain why $\log_b \dfrac{M}{N} = \log_b M - \log_b N$.

Practice

Write the expression as a single logarithm.

2. $\log_5 10 + 2\log_5 4.8$

3. $3\log_z \dfrac{1}{2} - \log_z 4$

Write the logarithm as a sum or difference of logarithms.

4. $\log_3\left(x^a r^3 f^n\right)$

5. $\log_c\left(\dfrac{x^3 p^9}{t^5}\right)$

Solving Logarithmic and Exponential Equations
Topic C.2 Common and Natural Logarithms

Vocabulary

common logarithm   •   natural logarithm   •   base

1. For all real numbers $x > 0$, the _____ of x is $\log_{10} x$, which is often written as $\log x$.

2. For all real numbers $x > 0$, the _____ of x is $\ln x = \log_e x$.

Step-by-Step Video Notes
Watch the Step-by-Step Video lesson and complete the examples below.

Example	Notes
1–3. On a scientific or graphing calculator, approximate the following values. $\log 7.32$ $\log 73.2$ $\log 0.314$	

Example	Notes
4–6. On a scientific or graphing calculator, approximate the following values.	
$\ln 7.21$	
$\ln 72.1$	
$\ln 0.0356$	

Helpful Hints

The number e is an irrational number that occurs in many formulas that describe real-world phenomena.

The logarithm, base b, of a positive number, x, is the power (exponent) to which the base b must be raised to produce x and is represented by the expression $\log_b x$.

Concept Check

1. On a scientific or graphing calculator, approximate $\log 2.14$ and $\dfrac{\ln 2.14}{\ln 10}$ to the nearest thousandth. What do you notice?

Practice

On a scientific or graphing calculator, approximate $\log x$ and $\ln x$ for the following values of x. Round to the nearest thousandth as needed.

2. $x = 3.9658$ 4. $x = 100$

3. $x = 1456$ 5. $x = e^{4.286}$

Name: _____   Date: _____

Instructor: _____   Section: _____

Solving Logarithmic and Exponential Equations
Topic C.3 Change of Base of Logarithms

Vocabulary
change of base formula   •   common logarithm   •   natural logarithm

1. The _____ states that $\log_b x = \dfrac{\log_a x}{\log_a b}$, where a, b, and $x > 0$, $a \neq 1$, and $b \neq 1$.

Step-by-Step Video Notes
Watch the Step-by-Step Video lesson and complete the examples below.

Example	Notes
1. Evaluate the logarithm using common logarithms. Round to three decimal places. $\log_3 11$ Answer:	
2 & 3. Evaluate each logarithm using common logarithms. Round to three decimal places. $\log_{15} 12$ $\log_7 5.12$	

411

Example	Notes
4. Evaluate the logarithm using natural logarithms. Round to three decimal places. $\log_4 0.005739$ Answer:	
5 & 6. Evaluate each logarithm using natural logarithms. Round to three decimal places. $\log_{21} 436$ $\log_6 0.315$	

Helpful Hints

It is possible to evaluate $\log_b x$ by determining the power to which b must be raised to produce x, but when x is not a power of b, the change of base formula can be used to approximate the value of $\log_b x$.

To check your answer, y, for the approximation of $\log_b x$, evaluate b^y and check that it is equal to x. Note that when dealing with approximated answers, you should check that b^y is approximately equal to x.

Concept Check

1. Explain why the change of base formula is necessary to compute $\log_2 7$ but not to compute $\log_2 0.125$.

Practice

Evaluate each logarithm using common logarithms. Round to three decimal places.

2. $\log_7 147$

3. $\log_{100} 0.01$

Evaluate each logarithm using natural logarithms. Round to three decimal places.

4. $\log_{14} 8.24$

5. $\log_{4982} 568.2569$

Solving Logarithmic and Exponential Equations
Topic C.4 Solving Exponential Equations and Applications

Vocabulary
taking the logarithm of both sides   •   growth equation   •   solving exponential equations

1. The _____ property states that if x and $y > 0$, and $x = y$, then
   $\log_b x = \log_b y$, where $b > 0$ and $b \neq 1$.

Step-by-Step Video Notes
Watch the Step-by-Step Video lesson and complete the examples below.

Example	Notes
1. Solve the exponential equation. Leave your answer in exact form. $2^x = 7$ Take the logarithm of each side of the equation. Solve the equation. Check the solution. Answer:	
3. Solve the exponential equation. Approximate your answer to the nearest ten-thousandth. $e^{2.5x} = 8.42$ Take the natural logarithm of each side of the equation. Answer:	

Example	Notes
4. If P dollars is invested in an account that earns interest at 12% compounded annually, the amount available after t years is $A = P(1+0.12)^t$. How many years will it take for $300 dollars in this account to grow to $1500? Round your answer to the nearest whole year.	

Answer:

Helpful Hints
To solve exponential equations, take the logarithm of each side of the equation. Then, solve the resulting equation using the properties of logarithms and check the solution.

The growth equation for populations that grow continuously is $A = A_0 e^{rt}$, where A is the amount at time, t, A_0 is the original amount, r is the rate at which things are growing in a unit of time, and t is the total number of time in years.

Concept Check
1. Solve $10^x = 1$ by taking the common logarithm of both sides. Then, solve it by converting it to a logarithmic equation. Do you get the same answer? Explain.

Practice
Solve the exponential equation. Approximate your answer to the nearest thousandth.

2. $e^{2.5x} = 19$

3. $4^{4x+1} = 28$

4. $110^x = 14.58$

5. If the population of mosquitoes in a small town is nine billion and it continues to grow at a rate of 3% per year, how many years will it take for the population to increase to fifteen billion mosquitoes? Round your answer to the nearest whole year.

Solving Logarithmic and Exponential Equations
Topic C.5 Solving Logarithmic Equations and Applications

Vocabulary
logarithmic property of equality   •   logarithmic equation

1. The _____ states that if $\log_b x = \log_b y$, then $x = y$, where $b > 0$ and $b \neq 1$, and x and y are positive real numbers.

Step-by-Step Video Notes
Watch the Step-by-Step Video lesson and complete the examples below.

Example	Notes
1. Solve $\log 5 = 2 - \log(x+3)$ for x. Isolate the logarithm. Convert the equation to an exponential equation. Solve the equation. Check the solution(s). Answer:	
3. Solve for x. $\log(x+6) + \log(x+2) = \log(x+20)$ Answer:	

Example	Notes
4. The magnitude of an earthquake is measured by the formula $R = \log\left(\dfrac{I}{I_0}\right)$, where I is the intensity of the earthquake and I_0 is the minimum measurable intensity. The 1964 earthquake in Anchorage, Alaska, had a magnitude of 8.4. The 1906 earthquake in Taiwan had a magnitude of 7.1. How many times more intense was the Anchorage earthquake than the Taiwan earthquake?	

Answer:

Helpful Hints

To solve logarithmic equations, if there is more than one logarithmic term, use the properties of logarithms to rewrite them as a single logarithm. Then, get one logarithmic term on one side of the equation and one numerical value on the other. Finally, convert the equation to an exponential equation and solve. Be sure to check any solutions.

To solve logarithmic equations with only logarithmic terms, use the properties of logarithms to write each side as a single logarithm. Then, use the logarithmic property of equality and solve the resulting equation. Be sure to check any solutions.

Concept Check

1. When solving the equation $\log_6(x-3) + \log_6(x+2) = 1$ for x, Paul found that $x = 4$ and $x = -3$. Explain Paul's error.

Practice

Solve for x.

2. $\log_4(x+3) + \log_4(x-3) = 2$

3. $\log_7(x+6) - \log_7(x+2) = \log_7(x+3)$

4. How many times more intense is a magnitude 5.5 earthquake than a magnitude 5.1 earthquake?

Additional Topics
Topic D.1 The Midpoint Formula

Vocabulary

midpoint   •   midpoint formula   •   length

1. The _____ of a line segment is the point on the line segment that is the same distance from each of the endpoints of the line segment.

Step-by-Step Video Notes
Watch the Step-by-Step Video lesson and complete the examples below.

Example	Notes

1. Find the midpoint of the line segment whose endpoints are $(1,4)$ and $(3,2)$.
 	
Let (x_1, y_1) represent one point and (x_2, y_2) represent the other.
 	
$(\Box, \Box)$
  x_1 y_1
 	
$(\Box, \Box)$
  x_2 y_2
 	
Substitute the values into the midpoint formula and simplify.
 	
$\dfrac{x_1 + x_2}{2} = \dfrac{\Box + \Box}{2} = \dfrac{\Box}{2} = \Box$
 	
$\dfrac{y_1 + y_2}{2} = \dfrac{\Box + \Box}{2} = \dfrac{\Box}{2} = \Box$
 	
Answer:	

Example	Notes
2–4. Find the midpoint of each line segment whose endpoints are the following points. $(-5,0)$ and $(9,-4)$ $\left(\dfrac{1}{2},-\dfrac{3}{4}\right)$ and $\left(-\dfrac{1}{2},\dfrac{1}{4}\right)$ $(20,-6.5)$ and $(-13,-7.3)$	

Helpful Hints

Do not confuse finding the midpoint of a line segment with finding the distance of a line segment.

The distance between two points is a length, whereas the midpoint of a line segment is a point on the line.

Concept Check

1. What is the relationship between the distance between an endpoint of a line segment and the midpoint of a line segment and the distance between the two endpoints of a line segment?

Practice

Find the midpoint of each line segment whose endpoints are the following points.

2. $(2,3)$ and $(4,5)$

4. $\left(\dfrac{2}{3},-\dfrac{2}{3}\right)$ and $\left(-\dfrac{2}{3},\dfrac{1}{3}\right)$

3. $(-7,0)$ and $(9,-6)$

5. $(-2,-3.4)$ and $(15,-7.2)$

Name: _____   Date: _____
Instructor: _____   Section: _____

Additional Topics
Topic D.2 Writing Equations of Parallel and Perpendicular Lines

Vocabulary

parallel lines   •   perpendicular lines   •   negative reciprocals

1. ___*parallel line*___ are lines that are always the same distance apart and have no point of intersection.

2. ___*perpendicular lines*___ are lines that intersect at a 90° angle.

Step-by-Step Video Notes
Watch the Step-by-Step Video lesson and complete the examples below.

Example	Notes
1. Find the slope of a line parallel to $y = \dfrac{3}{4}x + 2$. Find the slope of the line. $\dfrac{3}{4}$ Find the slope of a parallel line. $\dfrac{3}{4}$ Answer: $\dfrac{3}{4}$	$m_1 = m_2$ Slope is equal
2. Find the slope of a line perpendicular to $6x + 2y = 9$. Find the slope of the line. -3 Find the slope of a perpendicular line. Answer: 3	negative reciprocal

419

Example	Notes
4. Determine if the lines are parallel, perpendicular, or neither. $y = 2x + 3$ $-8x + 4y = -4$ $4y = 8x - 4$ $\dfrac{4y}{4} = \dfrac{8x}{4} - \dfrac{4}{4}$ $y = 2x - 1$ Answer: _parallel_	
7. Find the equation of a line perpendicular to $y = \dfrac{3}{5}x + 8$ that passes through the point $(3, -1)$. Write the answer in slope-intercept form. $y = \dfrac{-3}{5}x + 2$ Answer: $y = \dfrac{-5}{3}x + 4$	

Helpful Hints
The symbol for parallel lines is ||. The symbol for perpendicular lines is ⊥.

All horizontal lines are parallel to each other and all vertical lines are parallel to each other. Horizontal lines are perpendicular to vertical lines.

Concept Check
1. Is it possible for a set of two lines to be both parallel and perpendicular? Explain.

Practice
Find the slope of a line parallel to and a line perpendicular to the given line.

2. $5x + 3y = 12$

3. $y = 5x + 3.5$

4. Find the slope of a line containing the points $(-2, 3)$ and $(4, 3)$. Then find the slope of a line parallel to this line and the slope of a line perpendicular to this line.

Additional Topics
Topic D.3 Graphing Linear Inequalities in Two Variables

Vocabulary
linear inequality   •   test point   •   integer solution

1. When graphing a(n) _____, first replace the inequality symbol with an equality symbol. Then graph the line.

Step-by-Step Video Notes
Watch the Step-by-Step Video lesson and complete the examples below.

Example	Notes
1. Graph $4x + 3y \le 9$. Graph the line. $\left(\boxed{0},\boxed{3}\right)$ $\left(\boxed{3},\boxed{-1}\right)$ $\left(\boxed{0},\boxed{}\right)$ Test a point. $\left(\boxed{0},\boxed{0}\right)$ $4\left(\boxed{}\right) + 3\left(\boxed{}\right) \le 9$ *True* This is a _____ statement. Shade to show all solutions.	

Answer:

Example	Notes

2. Graph $3y < -2x$.

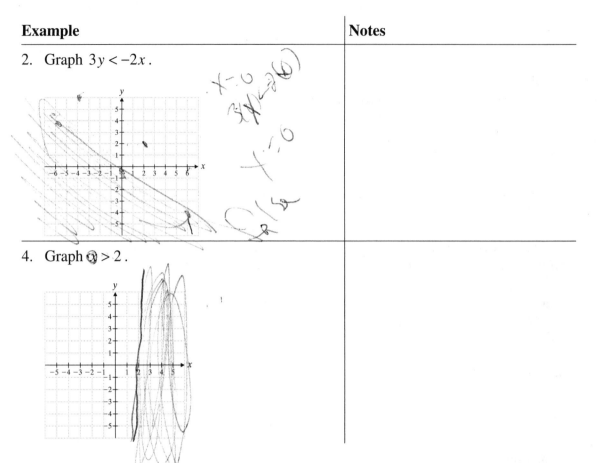

4. Graph $x > 2$.

Helpful Hints
The origin $(0,0)$ is usually the most convenient test point to pick. If the point $(0,0)$ is a point on the line, choose another convenient test point.

When the inequality symbol is $\leq$ or $\geq$, use a solid line when graphing the inequality. When the inequality symbol is $<$ or $>$, use a dotted line when graphing the inequality.

Concept Check
1. What test point would you choose to test the linear inequality $y > x$? Explain.

Practice
Graph the given inequality.

2. $y \geq 2x + 3$

3. $y \leq -2$

4. $x - y \leq 0$

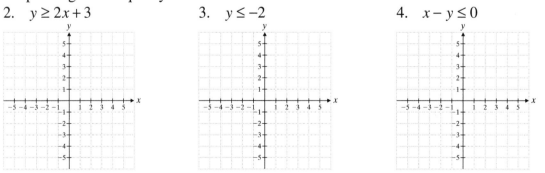

Additional Topics
Topic D.4 Systems of Linear Inequalities

Vocabulary
system of linear inequalities   •   test point   •   point of intersection

1. Two or more linear inequalities graphed on the same set of axes is called a
    _____.

Step-by-Step Video Notes
Watch the Step-by-Step Video lesson and complete the examples below.

Example	Notes
2. Graph the solution to the system of linear inequalities. $y < -\dfrac{4}{3}x + 3$ $y \geq \dfrac{1}{2}x - 2$	
3. Graph the solution to the system of linear inequalities. $y \geq 2x + 3$ $4x - 2y > 4$	

Example	Notes
4. Graph the solution to the system of linear inequalities. Find and label all the points of intersection. $x + y \leq 3$ $x - y \leq 1$ $x \geq -1$	

Helpful Hints
The solution to a system of linear inequalities is the intersection of the solution sets of the individual inequalities.

If the graph of a system of linear inequalities is parallel lines with the shading outside the lines, there is no solution to the system.

Concept Check
1. Can a system of two linear inequalities have no solution if it has two perpendicular lines?

Practice
Graph the solution to the given system of inequalities.

2. $y \geq 0.5x$
   $x \geq 2y + 3$

3. $y \geq 3$
   $x < -1$

4. $y - x > 2$
   $y < 2$
   $y \geq -2x - 4$

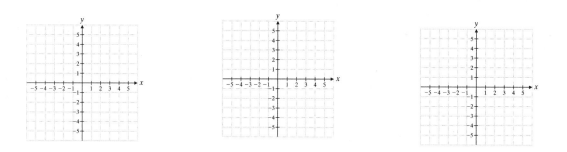

Additional Topics
Topic D.5 Synthetic Division

Vocabulary
synthetic division   •   quotient   •   divisor   •   dividend

1. Dividing a polynomial by a binomial can be made more efficient with a process called
   _____.

Step-by-Step Video Notes
Watch the Step-by-Step Video lesson and complete the examples below.

Example	Notes
1. Divide using synthetic division. $\left(3x^3 + 7x^2 - 4x + 3\right) \div \left(x + 3\right)$ Turn the first three numbers in the bottom row into the quotient. The last number in the bottom row is the remainder. Put this over the divisor to complete the answer. Answer:	

Example	Notes
2. Divide using synthetic division. $\left(3x^4 - 21x^3 + 31x^2 - 25\right) \div \left(x - 5\right)$	

Helpful Hints

"Missing terms" refers to monomial terms with degrees between the degrees of the terms in the polynomial being divided. For example, if the polynomial being divided is $3x^3 + 1$, the missing terms are $0x^2$ and $0x$.

Be sure to include zeros for the missing terms when using synthetic division.

Concept Check
1. What is the disadvantage of synthetic division?

Practice
Divide using synthetic division.

2. $\left(x^3 + 4x + 4\right) \div \left(x + 1\right)$

3. $\left(x^5 + 4x^2 + x\right) \div \left(x - 2\right)$

4. $\left(x^4 + 3x^3 + 5x^2 + 18x + 9\right) \div \left(x + 3\right)$

5. $\left(x^3 - 8\right) \div \left(x - 2\right)$

Name: _Sydney_
Instructor: _schnabl_

Date: _3/7/16_
Section: _____

Additional Topics
Topic D.6 Complex Numbers

Vocabulary

imaginary numbers   •   complex numbers   •   powers of i

1. The set of _imaginary numbers_ consists of numbers of the form bi where b is a real number, and $b \neq 0$.

2. _Complex numbers_ are numbers of the form $a + bi$, where a and b are real numbers.

Step-by-Step Video Notes
Watch the Step-by-Step Video lesson and complete the examples below.

Example	Notes
1 & 2. Simplify. $\sqrt{-49} = \sqrt{\boxed{2}} \cdot \sqrt{\boxed{9}} = \boxed{?} = \boxed{?}\, 7i$ $\sqrt{-8}$ $i\, 2\sqrt{2}$ $2i\sqrt{2}$	$\sqrt{-49} = \sqrt{-1} \cdot \sqrt{49} = 17 \;(7i)$ $\sqrt{-8} = \sqrt{4 \cdot 2} \cdot \sqrt{-1} = (2i\sqrt{2})$
3 & 4. Evaluate the powers of i. $i^9 = \left(\boxed{?}\right)\left(\boxed{?}\right)\left(\boxed{?}\right) = \left(\boxed{}\right)\left(\boxed{}\right)\left(\boxed{i}\right) = \boxed{i}$ $i^{15}\quad i^4 i^4 i^4 i^3 \quad -1 \cdot 1 \quad -1$	$i^9 = i^4 \cdot i^4 \cdot i$ $= 1 \cdot 1 \cdot i$ $= i$ $i^{15} = i^4 \cdot i^4 \cdot i^4 \cdot i^3$ $1 \cdot 1 \cdot 1 \cdot i^3 \;(-i)$
5. Add or subtract. $(5 + 6i) + (6 - 3i) = \boxed{5} + \boxed{6} + \boxed{6i} - \boxed{3i}$ $= \left(\boxed{5} + \boxed{6}\right) + \left(\boxed{6i} - \boxed{3i}\right)$ $= \boxed{11} + \boxed{3i}$ Answer: $11 + 3i$	$(5 + 6) + (6i - 3i)$ $\boxed{11 + 3i}$

Example	Notes

8. Multiply.

$(7 - 6i)(2 + 3i)$

Answer: $5 - 3i$

$(7-6i)(2+3i)$

$14 + 7(3i) + 2(-6i) - 6i(3i)$

$14 + 21i - 12i - 18i^2$

$14 + 21i - 12i - 18(-1)$

$\boxed{32 + 9i}$

---

10. Divide.

$\dfrac{7+i}{3-2i}$

$\dfrac{7}{3} - \dfrac{1}{2i}$

$\dfrac{14+77i}{13+8i}$

Answer: $\dfrac{19}{13} + \dfrac{17i}{13}$

$\dfrac{7+i}{3-2i} \cdot \dfrac{3+2i}{3+2i}$ $\quad (3-2i)(3+2i)$

$9 + 6i - 6i - 4i^2$

$(7+i)(3+2i)$ $\quad 9 + 4 = 13$

$21 + 14i + 3i + 2i^2$ $\quad \boxed{\dfrac{19}{13}}$

$21 + 14i + 3i + 2(-1)$

$\boxed{19 + 17i}$

Helpful Hints

For all positive real numbers a, $\sqrt{-a} = \sqrt{-1} \cdot \sqrt{a} = i\sqrt{a}$.

Note that $i^{4n} = 1$ where $n \neq 0$. Keep this in mind when evaluating powers of i.

Concept Check

1. Why is $1 + \pi \cdot i^4$ a real number while $\pi + i^5$ is not? Are they complex numbers? If so, what are their real and imaginary parts?

Practice

Add, subtract, multiply, or divide as indicated.

2. $(1.28 + 2i) + (11 - 5i)$ 4. $(6 - 2i)(2 + 7i)$

3. $\left(\dfrac{3}{2} + 8i\right) - (3 - 6i)$ 5. $\dfrac{1 + 4i}{7 - 3i}$

Name: _____   Date: _____

Instructor: _____   Section: _____

Additional Topics
Topic D.7 Matrices and Determinants

Vocabulary
matrix • dimensions of a matrix • determinant • value of a 2×2 determinant
value of a 3×3 determinant • minor of an element of a 3×3 determinant • array of signs

1. A(n) _____ is a rectangular array of numbers that is arranged in rows and columns.

2. The _____ $\begin{vmatrix} a & c \\ b & d \end{vmatrix}$ is $ad - bc$.

Step-by-Step Video Notes
Watch the Step-by-Step Video lesson and complete the examples below.

Example	Notes
1. Write the coefficients of the variables in the system of equations as a matrix. $5y = 10$ $2x - 6y = 17$ $\begin{bmatrix} \square & \square \\ \square & \square \end{bmatrix}$	
3. Find the determinant of the matrix. $\begin{bmatrix} 0 & -3 \\ -2 & 6 \end{bmatrix}$ Answer:	
4. Evaluate the determinant. $\begin{vmatrix} 4 & 1 & 2 \\ 3 & -1 & 0 \\ 1 & 2 & 3 \end{vmatrix}$ Answer:	

Example	Notes
7. Evaluate the determinant by expanding it by minors of elements in the first column. $$\begin{vmatrix} 2 & 3 & 6 \\ 4 & -2 & 0 \\ 1 & -5 & -3 \end{vmatrix}$$ Answer:	

Helpful Hints

To evaluate a 3×3 determinant, use expansion by minors of elements in the first column.

The minor of an element (number or variable) of a 3×3 determinant is the 2×2 determinant that remains after we delete the row and column in which the element appears.

Concept Check

1. What does the array of signs help to determine?

Practice

Evaluate the determinants.

2. $\begin{vmatrix} 4 & 0 \\ 10 & -1 \end{vmatrix}$

3. $\begin{vmatrix} 3 & 4 \\ -1 & -2 \end{vmatrix}$

Evaluate the determinant by expanding by minors.

4. $\begin{vmatrix} 1 & 3 & 4 \\ -2 & 0 & 0 \\ -4 & 5 & -1 \end{vmatrix}$

Additional Topics
Topic D.8 Cramer's Rule

Vocabulary
Cramer's rule   •   value of a 2×2 determinant   •   system of linear equations

1. _____ states that the solution to $\begin{matrix} a_1 x + b_1 y = c_1 \\ a_2 x + b_2 y = c_2 \end{matrix}$ is $x = \dfrac{D_x}{D}$ and $y = \dfrac{D_y}{D}$,

   $D \neq 0$, where $D = \begin{vmatrix} a_1 & b_1 \\ a_2 & b_2 \end{vmatrix}$, $D_x = \begin{vmatrix} c_1 & b_1 \\ c_2 & b_2 \end{vmatrix}$, and $D_y = \begin{vmatrix} a_1 & c_1 \\ a_2 & c_2 \end{vmatrix}$.

Step-by-Step Video Notes
Watch the Step-by-Step Video lesson and complete the examples below.

Example	Notes
1. Solve the system by Cramer's rule. $$-3x + y = 7$$ $$-4x - 3y = 5$$ Find D. $$D = \begin{vmatrix} \Box & \Box \\ \Box & \Box \end{vmatrix} = (\Box)(\Box) - (\Box)(\Box) = \Box$$ Find D_x. Find D_y. Solve for x and y. $$x = \frac{D_x}{D} = \frac{\Box}{\Box} = \Box \text{ and } y = \frac{D_y}{D} = \frac{\Box}{\Box} = \Box$$ Answer:	

Example	Notes
3. Solve the system by Cramer's rule. $2x - y + z = 6$ $3x + 2y - z = 5$ $2x + 3y - 2z = 1$ Answer:	

Helpful Hints

The solution to $\begin{matrix} a_1x + b_1y + c_1z = k_1 \\ a_2x + b_2y + c_2z = k_2 \\ a_3x + b_3y + c_3z = k_3 \end{matrix}$ is $x = \dfrac{D_x}{D}$, $y = \dfrac{D_y}{D}$, and $z = \dfrac{D_z}{D}$, $D \neq 0$, where

$$D = \begin{vmatrix} a_1 & b_1 & c_1 \\ a_2 & b_2 & c_2 \\ a_3 & b_3 & c_3 \end{vmatrix}, \ D_x = \begin{vmatrix} k_1 & b_1 & c_1 \\ k_2 & b_2 & c_2 \\ k_3 & b_3 & c_3 \end{vmatrix}, \ D_y = \begin{vmatrix} a_1 & k_1 & c_1 \\ a_2 & k_2 & c_2 \\ a_3 & k_3 & c_3 \end{vmatrix}, \text{ and } D_z = \begin{vmatrix} a_1 & b_1 & k_1 \\ a_2 & b_2 & k_2 \\ a_3 & b_3 & k_3 \end{vmatrix}.$$

The value of the 2×2 determinant $\begin{vmatrix} a & b \\ c & d \end{vmatrix}$ is $ad - bc$.

Concept Check

1. Would you solve the system $\begin{matrix} x + y = 1 \\ -x + y = 3 \end{matrix}$ using Cramer's rule? Explain why or why not.

Practice
Solve the systems by Cramer's rule.

2. $\quad 3x = 9$
$-8x + 10y = -4$

3. $4x + 6y = 48$
$-x + 2y = 16$

4. $\quad x + 2y + z = 5$
$-x + 4y - 2z = 12$
$-3x + y = 9$

Additional Topics
Topic D.9 Solving Systems of Linear Equations Using Matrices

Vocabulary
augmented matrix   •   matrix row operation

1. A matrix that is derived from a system of linear equations is called the _____ of the system.

Step-by-Step Video Notes
Watch the Step-by-Step Video lesson and complete the examples below.

Example	Notes
1. Solve the system of equations using a matrix. $4x - 3y = -13$ $x + 2y = 5$ Write the augmented matrix and perform row operations to solve the system of equations. Write the equivalent system and solve. Answer:	

Example	Notes
2. Solve the system of equations using a matrix. $$2x + 3y - z = 11$$ $$x + 2y + z = 12$$ $$3x - y + 2z = 5$$ Answer:	

Helpful Hints

An augmented matrix is made up of two smaller matrices separated by a vertical line. The coefficients of the variable terms are placed to the left of the vertical line, and the constant terms are placed to the right.

In a matrix, all the numbers in any row or multiple of a row may be added to the corresponding numbers of another row. All the numbers in a row may be multiplied or divided by any nonzero number. Any two rows of a matrix may be interchanged.

Concept Check
1. Can you think of a situation where solving a system of linear equations using matrices would not be the best method? Explain.

Practice
Solve the systems of equations using matrices.

2. $-2x + 3y = 2$
    $-4x + y = -16$

3. $4x - 2y = 7$
    $3x + 2y = 7$

4. $x + 4y - 5z = 4$
    $3x - 8y + 7z = 9$
    $2x + 9z = 17$

Additional Topics
Topic D.10 Basic Probability and Statistics

Vocabulary

probability   •   favorable outcomes   •   possible outcomes

1. _____ measure(s) the likelihood that a given event will occur.

Step-by-Step Video Notes
Watch the Step-by-Step Video lesson and complete the examples below.

Example	Notes
2 & 3. Find the following probabilities based on rolling a six-sided die. Find the probability of rolling a 4. $$P\left(\square\right) = \frac{\square}{\square}$$ Find the probability of rolling a 1 or a 3. $$P\left(\boxed{}\right) = \frac{\square}{\square} = \frac{\square}{\square}$$	
4. If a coin is tossed twice, find the probability of throwing tails twice in a row. Begin by drawing a tree diagram. Use the tree diagram to determine the probability. Answer:	

Example	Notes
6. If a die is tossed and then a coin is tossed, find the probability of rolling a 4 and then tossing a head. Answer:	
8. A bag contains 6 red marbles, 7 blue marbles, 2 black marbles, and 5 green marbles. Find the probability of picking a blue or black marble. Answer:	

Helpful Hints

Probability can also be defined as $\text{Probability} = \dfrac{\text{Favorable Outcomes}}{\text{Possible Outcomes}}$.

When multiple events occur, the probability of each event is multiplied to get the probability for all the events.

Concept Check
1. Explain why probabilities can only be between 0 and 1.

Practice

A die is tossed and then a coin is tossed. Find the following probabilities.

2. The die lands on an even number and the coin lands on heads.

3. The die lands on a multiple of 3 and the coin lands on heads or tails.

A bag contains 4 red marbles, 7 green marbles, and 3 white marbles.

4. Find the probability of picking a black marble.

5. Find the probability of picking a red, green, or white marble.

Name: _Sydney_ _____   Date: _2/11/16_ _____

Instructor: _____Schnabel_____   Section: _____

Additional Topics

Topic D.11 Applications of Systems of Linear Equations

Vocabulary

system of equations   •   matrix   •   word problem

1. To solve a __word problem__ using a __system of equations__, read the question and
   determine what information you will need to solve the problem.

Step-by-Step Video Notes

Watch the Step-by-Step Video lesson and complete the examples below.

Example	Notes
1. A movie theater sells tickets for $10 and bags of popcorn for $3. In a single Saturday night, the theater had $2375 in sales. The theater owner found that if he raised ticket prices to $11 and the popcorn prices to $4 and sold the same number of tickets and popcorn bags, the theater would make $2700. How many tickets were sold? How many bags of popcorn were sold?	$t =$ # of tickets $p =$ # of popcorn $\begin{cases} 10t + 3p = 2375 \\ 11t + 4p = 2700 \end{cases}$ $40t + 12p = 9500$ $-33t + 12p = 8100$ _____ $7t = 1400$ $\boxed{t = 200}$ $10(200) + 3p = 2375$ $2000 + 3p = 2375$ $3p = 375$ $\boxed{p = 125}$

$t =$ # of tickets
$p =$ number of popcorn

$40t + 12p = 9500$ $10t + 3p = 2375 \times 4$
$33t - 12p = 8100$ $11t + 4p = 2700 \times -3$

$7t = 1400$ $\boxed{t = 200}$ $2000 + 3p = 2375$
                                    -2000 -2000

Answer: $p = 125$

| 2. A boat travels 20 miles upstream, against the current, in 4 hours. The return trip, 20 miles downstream, with the current, takes only 2 hours. Find the speed of the boat in still water and the speed of the current. | $r = 20$ mile $(r+c)$ down
$t = 4$ hour $(r-c)$ up

$20 = r-c(4)$ $4r - 4c = 20$
$20 = 4r - 4c$ $2(2r+2c=20)$
                    $(+)\ 4r+4c = 40$
$20 = v+c(2)$ _____
$20 = 2r + 2c$ $8r = 60$
                    $\boxed{r = 7.5}$
$4(7.5) - 4c = 20$ |

$r =$ speed still
$c =$ with current
$(r+c)$ down stream
$(r-c)$ against stream

Answer:

$30 - 4c = 20$
$-4c = -10$ $\boxed{c = 2.5}$

437

Copyright © 2012 Pearson Education, Inc.

Example	Notes

3. An electronics company makes two types of switches. Type A takes 4 minutes to make and requires $3 worth of materials. Type B takes 5 minutes to make and requires $5 of materials. In the latest production batch, it took 35 hours to make these switches, and the materials cost $1900. How many of each type of switch were made?

Handwritten work (left):

$A = A$
$B = B$

A 4m $3
B 5m 5

$S = $ # of switches
$D = $ amount paid

$4A + 5B = 2100$
$3A + 5B = 1900 \; (-1)$
$= 200$
A

$800 + 5B = 2100$
260

Answer: $A = 200$
$B = 260$

Handwritten work (right, Notes):

type $A = 4$ min
$+3$

type $B = 5$ min
$+5$

$4A + 5B = 2100$
$(-1)\; 3A + 5B = 1900$
_____
$\boxed{A = 200}$

$4(200) + 5B = 2100$
$800 + 5B = 2100$
$5B = 1300$
$\boxed{B = 260}$

Helpful Hints

Remember that once you solve a system relating to an application, it is important to go back and answer the question asked. This requires interpreting the solution to the system.

When you produce a system relating to an application, you can solve it using any method. Choose the method that is easiest for you or that is easiest to use with the particular system that you found.

Concept Check

1. While calculating the answer to Example 2, Tia incorrectly finds that the speed of the current is 0 mph and the speed of the boat is 20 mph. Explain her error.

Practice

2. A plane travels 3150 miles with the wind, in 7 hours. The return trip, 3150 miles against the wind, takes 9 hours. What is the speed of the plane?

4. Jim charges $40 to mow lawns and it takes him 3 hours to mow each one. He also charges $15 to weed gardens, which takes him an hour. In a month Jim worked 160 hours and made $2175. How many lawns did Jim mow?

3. What is the speed of the wind in Practice 2?

5. How many gardens did Jim weed in Practice 4?